Praise from the Experts

"*JMP Essentials* provides a simple, practical approach to problem solving. It will enable you to use JMP to solve complex business problems, even if you have no knowledge of statistics. *JMP Essentials* takes the scariness out of statistical terms and helps you make sense of your data and get answers to your questions. The one- or two-page format for each topic makes using the book simple, since it eliminates turning pages while working on your computer. Most important, this book will enable you to share your findings with others, which is essential to the process of getting support to improve your business."

Michelle Swart
Business Development Manager
Cisco Systems, Inc.

"I've used JMP for close to 20 years now, but I'm always learning new things. I knew the data filter existed, but I used a clumsier, home-developed method for doing what I wanted. I followed Curt and Chuck as they guided me through the data filter. I was impressed with the explanation. It was easy to follow, and I learned how to do what I wanted with a first pass through the discussion. I can only imagine that the novice will find similar success in acquiring the basic skills using their book. The layout helps the exposition because nearly every topic is covered in a page or in two facing pages, making learning the techniques quick and easy. I look forward to having my students use the guide. Not only will it save me that introductory hour or two to get them started, but after reading the entire book, they may be teaching me new tricks."

Richard De Veaux
Professor of Statistics
Williams College

"I know from personal experience that teaching statistics is easy. The hard part is getting people to understand the stuff, and getting them to actually use it is next to impossible. JMP, the statistical exploration tool from SAS, is a remarkable package that connects the seat of the intellect to the seat of the pants, allowing non-statisticians to quickly gain statistical insights in an interactive visual environment.

"However, the program has so many features that it can be intimidating for the beginner to dive into. There are excellent tutorials and a help system, but personally I find it more satisfying to pick up a book, and turn its pages. *JMP Essentials*, by Curt Hinrichs and Chuck Boiler, is the perfect introduction to JMP. And speaking of turning pages, they don't go from right to left, in this book, but from top to bottom. With its horizontal spiral binding, it was designed to sit flat, next to your keyboard, and help you with specific tasks.

"Clear step-by-step instructions get you up and running with JMP, describe its various data types, and help you create a wide variety of graphs. The information density was just right for me. The material is packed tightly enough so you can often stay on a single page to complete a task, yet it is not so dense that you need to go back and read things twice. I wish I had this book when I started using JMP. I'm glad I have it now."

Sam L. Savage
Author of *The Flaw of Averages*
Consulting Professor
Stanford University

Publishing

JMP® Essentials

An Illustrated Step-by-Step Guide for New Users

Curt Hinrichs

Chuck Boiler

The correct bibliographic citation for this manual is as follows: Hinrichs, Curt, and Boiler, Chuck. 2010. *JMP® Essentials: An Illustrated Step-by-Step Guide for New Users.* Cary, NC: SAS Institute Inc.

JMP® Essentials: An Illustrated Step-by-Step Guide for New Users

SAS Institute Inc., SAS Campus Drive, Cary, North Carolina 27513-2414.

June 2010

SAS provides a complete selection of books and electronic products to help customers use SAS® software to its fullest potential. For more information about our offerings, visit **support.sas.com/bookstore** or call 1-800-727-3228.

Contents

Preface

JMP Essentials was written for the new or occasional user of JMP software who needs to get the right results right away. If you have data and problems to solve, JMP will help you make sense of and understand them to arrive at a good decision. It is the goal of this book to help you with this pursuit.

We often find new users of JMP simply trying to complete a specific task with their data. Perhaps you just need to generate a certain graph for a Microsoft PowerPoint presentation or to quickly see how patterns in your data will lead you to an important discovery. If these scenarios sound close to home, you've come to the right place. This book is task oriented and will help you access your data, identify that graph or statistic you need, and quickly and easily create and share it with others.

Though JMP contains state-of-the-art visualization and advanced statistical and data mining tools, we also believe it is the right tool for the user who might be less confident in his or her statistical abilities. While there is a time and place for every feature in JMP, we have tried to include only those topics that a typical new user—a manager or analyst, for example—would need. If it's been a while since you've studied statistics, or dealt with statistical terminology, do not worry. We will present the key ideas and fill in the gaps as needed.

JMP software is built around the workflow of the problem solver. One of its outstanding features is that it consistently provides the correct graphs and statistics for the data you're working with (something we refer to as JMP's *smart interface*). JMP will lead you down the right path to the right result, provided your data is properly classified and you have an idea of what questions you're trying to answer.

JMP is easy to use. In most cases, generating a graph or result will take you seconds, maybe minutes, to complete but never hours or days. Much of this efficiency is due not only to the small number of steps required but to the ability to navigate intuitively toward the right solution quickly, rather than through repetitive and time-consuming trial and error. This book provides you with the *essential* knowledge to get to your solution even faster.

Audience

JMP Essentials was written for the new or occasional user of JMP. We have focused on the most commonly used features and we have provided the needed instruction to generate results quickly. Each key step in this process is illustrated with screen shots to help you see the result and develop your confidence using JMP. We don't assume any formal background in statistics. Instead we emphasize the intuition of concepts over statistical theory. If you require deeper statistical understanding, we recommend some excellent textbooks in the Bibliography.

Most new JMP users have one of the following distinct needs:

1. They have a good idea of what graph they need and simply want to create it (see Chapter 3).

2. They really don't know what the data will say and need help exploring or summarizing it (see Chapter 4).

3. They need to make sense of, or answer some specific questions about, the data they have (see Chapters 5 and 6).

We also believe that the complete book is well suited as a reference guide to the following groups of users:

- **Spreadsheet users** who are looking for a convenient way to produce nice visualizations of their data or to supplement a spreadsheet's statistical capabilities. JMP reads and writes data from a variety of programs including Microsoft Excel. This book provides a quick and easy way to make your spreadsheet data come alive and allows you to fully and interactively explore that data.

- **Students enrolled in introductory statistics courses** who need JMP instruction. JMP is the ideal tool for students because its navigation reinforces the basic assumptions taught in an introductory course. This book provides an overview of the JMP tools needed in most first-year courses.

- **SAS users** who want to take advantage of JMP's data visualization tools. JMP integrates beautifully with SAS, and we've provided Appendix A to illustrate some of these features.

Approach and Features

We have found that the best way to learn JMP is by using it and getting value out of it quickly. Our goal is to present the materials in this book in the most user-oriented approach possible. So, we have made every effort to organize the presentation around the new user's common needs and questions and the most direct and concise means to answer them. We also recognize that the most basic use of data is in generating graphs of data rather than performing more complex statistical analyses. The following features are included for this purpose:

- We present the material with a **show-and-tell** approach. In most cases, we show you what the results look like alongside the conditions and steps required to produce them. We think this approach is especially useful for JMP users who have a good idea of what they want from JMP and just need the steps to create it.

- When appropriate, we provide an **example-driven context** for each JMP platform that explains its use, value, and general application to problems. We have tried to distill these contexts down to typical or easy-to-understand cases.

- We organize the contents into **easily manageable chunks of information**. While the entire book is designed to cover a fairly complete overview of the basics, each chapter or tab represents one family of tasks (such as importing data, creating graphs, and sharing graphs).

- We hope you will keep this book near your computer. Within each chapter or tab, we have designed each page—whenever possible—to be self-contained (beginning and ending a task in a page or page spread), allowing you to quickly find and execute the required steps you're looking for without having to flip pages.

- No matter what your professional background, this book assumes only that you have a basic working knowledge of Microsoft **Windows**. Virtually all of the information in this book applies to using JMP on the **Macintosh** operating system, but only the Windows version of JMP is used in the examples.

Organization

This book is designed like a cookbook. Find what you need, and follow the steps. We have organized the contents of this book to reflect both the process of analyzing data (getting data, analyzing it, and sharing the results) and the progression from the very basic features in JMP to more specialized ones. We hope this organization offers the most value to the reader. Much of our judgment in this regard comes from our experience working one-on-one with new JMP users.

- Chapter 1 covers the preliminary material you'll need for the rest of the book. The chapter identifies the conventions we use and introduces you to JMP menus, windows, and preferences.

- Chapter 2 covers the first step in any analysis: getting your data into JMP or creating it. With the exception of some material in Chapter 2, other chapters are self-contained, and you can read them in any order.

- Chapters 3 through 6 cover graphing and analysis:

  - Chapter 3 is for the user who knows what graph he or she wants.

  - Chapter 4 is for the user who does not know what the data says and needs to explore it to find an appropriate graph or summary.

- Chapters 5 and 6 are for the user who needs to solve a problem and answer questions using analytics and graphs.

- Chapter 7 covers topics related to sharing your graphs or results in a presentation or other document.

- Chapter 8 covers additional resources that are available within JMP, online and from outside resources, such as training, books, and user groups.

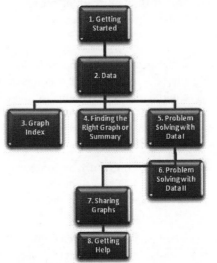

Figure 1 Chapter Organization

Authors' Web Site

We have created an authors' Web site for the book that provides updates, corrections, enhancements, and additional examples for new releases and offers a means to contact us with any suggestions or comments you have. We also provide links to helpful resources outlined in Chapter 8. Go to **support.sas.com/authors**.

We'd love to hear from you.

Acknowledgments

We'd first like to thank John Sall for creating a great product in JMP. Who knew that statistics could be so fun? If you are new to JMP, we hope it inspires you as it has us. Thanks also to Jon Weisz, Dave Richardson and John Leary for supporting the idea for this book.

We have been very fortunate to work with the outstanding professionals at SAS Press. Shelley Sessoms got us to the starting line, and Stephenie Joyner has been a constant source of advice and support, cheering us on throughout the development process. Thanks also to Julie Platt, Kathy Restivo, Mary Beth Steinbach, Monica McClain, Candy Farrell, Jennifer Dilley, Stacey Hamilton, and Shelly Goodin, at SAS Publishing for their encouragement.

The manuscript for this book was substantially improved by the insightful suggestions of our reviewers. The book contains many new features and refinements due to their input. Our reviewers include:

Chris Albright, Indiana University
Mark Bailey, SAS Institute
Peter Bruce, Statistics.com
Bob Lamphier, SAS Institute
Paul Marovich, SAS Institute
Gail Massari, SAS Institute

Tonya Mauldin, SAS Institute
Don McCormack, SAS Institute
Chris Olsen, Thomas Jefferson High School
Jeff Perkinson, SAS Institute
Lori Rothenberg, North Carolina State University
Heath Rushing, SAS Institute
Mia Stephens, SAS Institute
Sue Walsh, SAS Institute
Annie Zangi, SAS Institute

We also wish to thank Sam Savage of Stanford University for testing the book in his class, and David Shultz and Mary Loveless for using the book with customer training.

This book began with a desire to help the new JMP user and evolved into a labor of love. But without the love and incredible support of our families, friends and colleagues this project would have never materialized. Thank you!

Curt Hinrichs
Chuck Boiler

San Francisco, CA
June 2010

Chapter 1 • Getting Started

JMP was developed to help people with questions about their data get the answers they need through the use of graphs and numerical results. For most people, memories of statistics are a very unpleasant, if not forgotten, part of their education. If you see yourself as a new, occasional, or even reluctant user of data analysis, we want you to know that we have written this book for you.

It is important to note that throughout the historical development of statistics as a scientific discipline, people had real problems they needed to solve and developed statistical techniques to help solve them. Statistics can be thought of as sophisticated common sense, and JMP takes a practical, common sense approach to solving data-driven problems.

JMP is the ideal tool for those who need to make good sense of data because it was designed around the workflow of the data analyst rather than as a collection of tools only a statistician can understand. When you think about your data analysis problem, try to formulate the questions that might help you address it. For example, do you need to describe the variation in selling prices of homes in a city or understand the relationship of customer satisfaction with service waiting times? With this mindset, you will find the menus and navigation in JMP to be very compatible with the types of questions you are trying to answer.

Displaying graphs (or pictures) of data is one of JMP's strengths. For most people, an effective graph can convey more information more quickly than a table of numbers or statistics. In any JMP analysis, graphs are presented first and then the appropriate numerical results follow. This is by design. JMP also provides a **Graph** menu that contains additional visualization tools that are independent of numerical results. The goal of this chapter is to introduce you to JMP and its basic navigation. We cover the menus and windows and introduce you to the conventions used throughout the book.

Getting Started

1.1 Using *JMP Essentials*

All but one chapter in this book (Chapter 3, "Index of Graphs") is laid out in a consistent manner to help you generate results quickly. The format of the book has been designed to be used alongside your computer. After an introduction to the concept, we have designed each page or two-page spread to be self-contained. That is, with few exceptions, the steps required to produce a result begin and end without having to flip through pages.

We provide numbered steps on the left side of the page that generate the result illustrated on the right side of the page (see Figure 1.1).

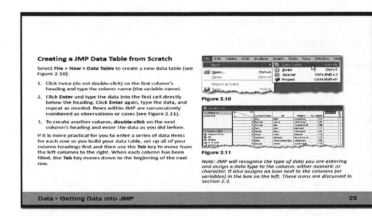

Figure 1.1

Note: This edition of JMP Essentials was written with JMP 8. However, the methods covered in this book are mostly basic and have not substantially changed since the earliest releases of the software. Thus, you will find most instructions contained in this book compatible with earlier and future JMP releases. If you are using JMP 9, you will find the instructions the same except where noted.

Conventions

We are confident that, having made it this far, you know the basic terminology associated with operating a computer, including click, right-click, double-click, drag, select, copy, and paste. We use these terms and they appear in numbered steps (see Figure 1.2). When there is a single or self-evident step, these instructions are included in the body of the text. Each step or action appears in bold type.

In writing this book, we have adopted the same conventions contained in JMP documentation to make your transition to using the documentation easy.

Menu items such as **Graph** are associated with a JMP command such as **Chart**. We use the greater than (>) symbol to indicate the next step in an operation. Thus, **Graph > Chart** indicates that you should select the **Chart** command (or platform) from the **Graph** menu (see Figure 1.3).

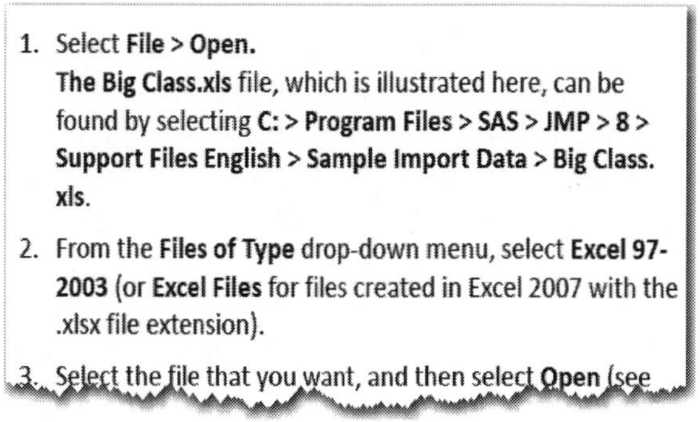

1. Select **File > Open**.
   The **Big Class.xls** file, which is illustrated here, can be found by selecting **C: > Program Files > SAS > JMP > 8 > Support Files English > Sample Import Data > Big Class. xls**.

2. From the **Files of Type** drop-down menu, select **Excel 97- 2003** (or **Excel Files** for files created in Excel 2007 with the .xlsx file extension).

3. Select the file that you want, and then select **Open** (see

Figure 1.2

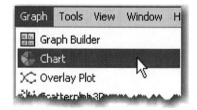

Figure 1.3

Getting Started

Book Features

Most chapters feature one or more examples to illustrate the procedures within that chapter (see Figure 1.4). All of the examples have corresponding data tables that are included in JMP's built-in Sample Data directory (**Help > Sample Data**).

Example 2.2 Movies

We will use the **Movies.jmp** data table to illustrate the concepts in this section. This data table consists of the 277 top-grossing movies released between 1937 and 2003. The columns are Movie, Type (genre), Rating (movie rating), Year (year in which the movie was released), Domestic$ (in millions of US dollars), Worldwide$ (in millions of US dollars), and Director. You can access this data table in the **Sample Data** folder that is installed with JMP by selecting **File > Open > C: > Program Files > SAS > JMP > 8 > Support Files English > Sample Data > Movies.jmp**.

Figure 1.4

Important definitions are boxed for easy reference (see Figure 1.5).

Data refers to any values placed in the cell of a JMP data grid. Examples include numeric and/or text descriptions: 3.6, $2500, Female, somewhat likely, or 9/24/09.

Data type refers to the nature of the data. The data type can be either numeric (numbers) or character (often words and letters but sometimes also numbers). The key difference is whether to treat the data as a number (numeric) or a category (character).

Modeling type refers to how the data within a column should be used in an analysis or a graph. JMP uses three distinct modeling types: continuous, nominal, and ordinal.

Figure 1.5

We include notes, tips, and cautions where appropriate to point out relevant or important information (see Figure 1.6).

Note: Once you've selected a new value, you can replace that value in the same column, create a new column with these values, or even create a formula column. Be careful! If you select **in Place**, these values cannot be changed because the Recode command replaces values in that column.

Figure 1.6

Where appropriate, we provide tables that summarize key information or translate jargon into common terms (see Figure 1.7).

What you want...	What that's called in JMP...	How to do it...
I want a bar chart.	Chart	**Graph > Chart**. Choose the measured column and add it to the **Statistics** area and choose a function. If you are not sure, choose **sum**, then click **OK**.

Figure 1.7

1.2 Launching JMP

Let's begin by launching JMP. To launch JMP from the
Microsoft Windows Start menu:

1. Select the **Start** menu.

2. Select **Programs**.

3. Select **JMP 8 > JMP 8** (see Figure 1.8).

Figure 1.8

Macintosh users can click on the JMP icon
(see Figure 1.9) to launch JMP from the
application dock.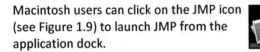

After JMP has launched, you might notice that two windows
have also opened, Tip of the Day and JMP Starter.

Figure 1.9

Tip of the Day

The Tip of the Day window is the first thing you see because it addresses the most common questions that new users ask, such as "How do I do *X*?" Well the *X* in these common questions is represented and answered in 36 different Tip of the Day windows. You can scroll through them by clicking **Next Tip** at the bottom of the window (Figure 1.10). Some of the Tip boxes contain important and basic navigational hints, while others only apply to more advanced features in JMP.

Note also the **Enter Beginner's Tutorial** button. This tutorial walks you through a basic analysis of data from opening a data table to creating graphs and results. JMP contains several other tutorials that are directed toward more specific types of problems and are found in the **Help** menu.

*Note: If you do not want to see the Tip of the Day window every time you launch JMP, you can simply uncheck the **Show tips at startup** box in the lower left corner of the window.*

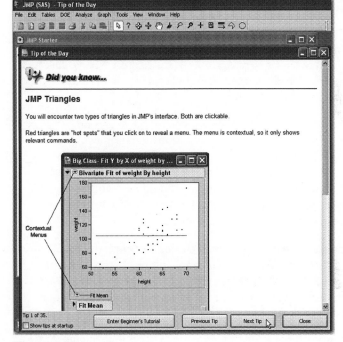

Figure 1.10

JMP Starter

After you close the Tip of the Day window, you see the JMP Starter window. The JMP Starter window is an optional and alternate means to navigate JMP (see Figure 1.11). It duplicates the functionality contained in JMP's native menus that are used throughout this book.

The JMP Starter feature was added to JMP in version 4 to assist those who are comfortable with statistical terminology or who might be transitioning from a competing product. It organizes JMP's features into general categories or families of statistical tools and provides direct links to those tools within a category. A nice feature of the JMP Starter is that it also describes the specific features within those links.

For example, if you click on the category **Graph** (see Figure 1.12), you see a selection of graph buttons to the right with a description of what they produce. Note that if you click on a button at this stage, you are prompted to open a data table (opening data tables is discussed later).

Note: This book focuses on using JMP's native menus (those that appear at the top of the JMP window). If you do not want the JMP Starter window to appear at start up, select **File > Preferences > General;** *uncheck the* **Initial JMP Starter Window** *box.*

Figure 1.11

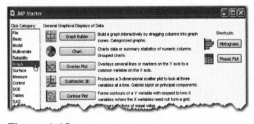

Figure 1.12

1.3 JMP Menus

At the top of the JMP window, you see a series of menus (**File**, **Edit**, **Tables**, and so on). These are the menus we use to illustrate the concepts in this book. They are also the same menus we refer to as JMP's native menus because they have been present in JMP since its first release.

These menus serve to open or import data, to edit or structure it, and to create graphs and analyses of your data. They are also a valuable source for assistance through the **Help** menu, which is discussed later. The menus are organized in a logical sequence from left to right.

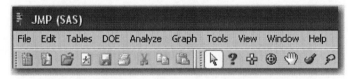

- **File** is where you go to open or import data and to save, print, or exit JMP. It is also where you can customize the appearance or settings within JMP through **Preferences** (explained in Section 1.5).

- **Edit** provides the usual cut, clear, copy, paste, and select functions, as well as undo and redo and special JMP functions.

- **Tables** provides the tools to manage, summarize, and structure your data (see Section 2.6).

- **DOE** contains the Design of Experiments tools, which we will not cover in this book. For more information, see **Help > Books > JMP DOE Guide**.

- **Analyze** contains the analysis tools that generate both graphs and statistics and serves as the home for all of JMP's statistical tools from simple to advanced (see Chapters 5 and 6).

- **Graph** contains graph tools that are independent of statistics (at least initially). Graphs in this menu include basic charts to advanced multivariable and animated visualization tools (see Chapters 3 and 4).

- **Tools** allows you to transform your mouse into a help tool, a brushing tool, a selection or scrolling tool, and much more (see Section 7.2).

- **View** provides options to navigate how you work with your files and windows.

- **Window** helps you manage windows within JMP.

- **Help** provides resources for learning and using JMP. Let's start with an introduction to the **Help** menu.

The Help Menu

The **Help** menu (see Figure 1.13) provides access to the learning resources you can use as you expand your knowledge of JMP and its features and learn about statistics and how to interpret results. These resources include indexes, guided tutorials, tips of the day, and searchable as well as printable books including the *JMP User Guide*. Data tables employed in this book and in all JMP documentation are included in the Sample Data directory. We cover the features of the **Help** menu in greater detail in Chapter 8.

JMP also features context-specific help, meaning that when you use the JMP Help Tool* in any graph or statistical result, you are directed to the right spot in the documentation to assist you in understanding the result. In statistical results, JMP provides Hover Help* that provides context-specific interpretations of certain statistics.

For descriptive graphs or basic summary statistics, interpretation can be straightforward, but as you dig deeper into an analysis or employ more advanced methods it is vitally important that you understand what the results mean, particularly when they are shared or presented. The documentation under **Help > Books** *includes over 2,700 pages of reference material in five books that address the needs*

of professional statisticians and analysts. If you encounter results that you do not understand, however, we strongly recommend that you seek assistance from experienced data analysts.

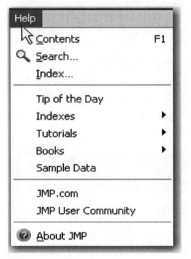

Figure 1.13

*The JMP Help Tool is discussed in Section 8.2. Hover Help is employed in statistical results and provides a context-sensitive interpretation, which is illustrated in Chapter 5.

The Analyze and Graph Menus

Because most graphs or statistical results begin with the **Analyze** and **Graph** menus, let's explore the structure within these two menus a little bit more here.

Click on the **Analyze** menu at the top of the window. Glance at the choices on the menu. Next, click on the **Graph** menu at the top of the window. Glance at the graph choices. The menus in JMP—specifically the **Analyze** and **Graph** menus (see Figures 1.14a and 1.14b)—are designed to provide both a description and visual cues for analyzing, graphing, and exploring data.

*Note that each entry under these menus has both a name and an icon. For graphical results, it is helpful to have an idea of what the final result will look like. The icons next to the **Graph** menu choices give you a preview of each graph. From the **Analyze** menu, the icons depict the description or relationships you will see in graphs and statistical results (Figure 1.15). You will see this menu item name and icon motif repeated throughout JMP's menus.*

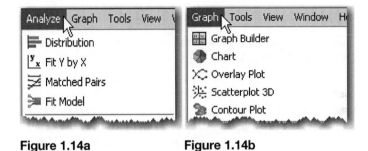

Figure 1.14a   **Figure 1.14b**

*Note: the **Analyze** menu items produce both graphs and statistical results while the **Graph** menu items produce only graphs.*

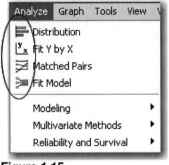

Figure 1.15

Getting Started

Framework of the Analyze Menu

There is a framework to the **Analyze** menu that we will discuss in detail in Chapter 5. As mentioned in the introduction, your exploratory objective will translate to these menu items. This structure streamlines the analysis process; you only need to count how many columns you have and know whether you are trying to describe, compare, or understand their relationship in order to select the correct menu item (see Figure 1.16).

This framework cues you to the correct analysis choice on the menu without exposing you to many statistical terms until you need it. Make no mistake, you still get the statistics when you want them, but you do not have to know all the statistical terms in order to access them.

*Note: JMP's **Analyze** menu contains terms such as Distribution and Fit Y by X that might be initially unfamiliar, but the ideas behind them are very straightforward. We describe them in simple terms as needed throughout the book. Many items under the **Analyze** and **Graph** menus are referred to as platforms through this book. For example, Distribution and Fit Y by X are referred to as platforms.*

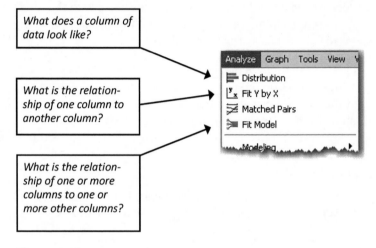

What does a column of data look like?

What is the relationship of one column to another column?

What is the relationship of one or more columns to one or more other columns?

Figure 1.16

1.4 JMP Windows

Throughout this book, each set of instructions used to create a graph or an analysis is prompted by a standardized window that follows a consistent format and execution. To launch a window, however, you must first open a data table.

Figure 1.17

For purposes of illustration, we will open the Equity.jmp data table:

1. Select **Help > Sample Data > Open the Sample Data Directory > Equity.jmp**.

2. Select **Analyze > Distribution** (see Figure 1.17).

3. This generates the Distribution window with the columns (variables) from the Equity.jmp data table populated under the Select Columns window (see Figure 1.18).

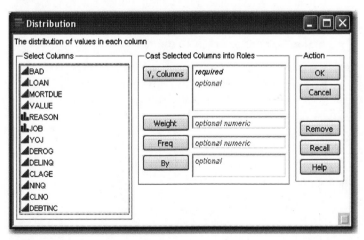

Figure 1.18

Getting Started

Most JMP windows consist of three main elements, organized from left to right (see Figure 1.19):

1. **Available columns** (or variables) of data to analyze from your data table. These appear on the left under **Select Columns**.

2. **Roles** that you want to place (or cast) on the column(s). In this area, you see buttons and empty areas under **Cast Selected Columns into Roles**. Within these empty areas, you are given a hint in italics about which columns are required and which are optional to run the analysis.

3. **Action buttons** to execute commands.

To use this Distribution window or almost any other in JMP, click on a column and select the role (or click and drag the column into that role). Once you are satisfied with your selections, select **OK**.

Almost every analysis and graph window in JMP appears in this format. So, when you learn this one window format, you learn just about every other one.

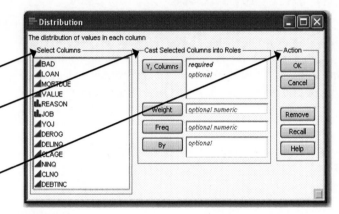

Figure 1.19

*Note: The **Y, Columns** role refers to what column you want to place on the vertical, or y, axis. In other windows, such as Fit Y by X, you also have an **X, Columns** role to select that corresponds to the horizontal, or x, axis.*

1.5 JMP Preferences

JMP's **Preferences** determine the way JMP appears or behaves on your machine. There are many preferences that can be adjusted. You can think of these as global settings. If you currently use JMP in a corporate environment, it's possible your JMP administrator has set these preferences to match a corporate standard. In this book, we introduce JMP using the default preferences that are set when JMP is installed. To view the preferences, choose **File > Preferences** (see Figure 1.20).

This generates the Preferences window (see Figure 1.21). Similar to the JMP Starter described earlier, the Preferences window has a panel containing 12 main categories on the left and options within those categories on the right. You can change preferences by checking or unchecking the boxes or by selecting items from drop-down menus. Changing preferences affects such things as the graph or result format, the font, the location of a file, and much more, each and every time you use those features in JMP. If you are unsure about making a change to the preferences, we recommend that you wait until you have a need to do so.

Figure 1.20

Figure 1.21

If you need to make a change within a single graph or result, note that JMP also provides many of these formatting options within the graphs themselves.

Getting Started

1.6 Summary

JMP was developed to help the business professional, scientist, or engineer get answers to the questions and problems they encounter. The navigation and menus within JMP provide a natural extension of your problem solving and a direct means to explore your data and generate the results you need. This book uncovers the structure of the JMP menus and provides easy steps for producing results. The standardized format of the windows in JMP prompts you through most analysis and graphing. Results can be customized using global detailed preferences.

Chapter 2 • Data

The first step in creating a graph or analysis is to get your data into JMP. With JMP, you can easily import data from many different sources such as Microsoft Excel or Microsoft Access, or you can enter your data directly into a JMP data table. Because most readers already have data in one form or another, the focus of this section is getting that data into JMP from another application. Sometimes data isn't in the best condition when you import it. Later in this chapter, we discuss what you can do to format data or deal with missing data.

As mentioned in the previous chapter, instructions throughout the book focus on using JMP's native menus (the menus at the top of your JMP window – see below). We also use Windows as our default system to illustrate JMP. JMP instructions for Windows and Macintosh are basically the same though some operating system differences are noted when they occur.

Data

Example 2.1 Big Class

We will be using the Big Class.jmp data file to illustrate the steps in this chapter. This data set consists of 40 middle-school students and their name, height, weight, gender, and age. You can access this data set in the Sample Data folder that is installed with JMP:

File > Open > C: > Program Files > SAS > JMP > 8 > Support Files English > Sample Data > Big Class.jmp

2.1 Getting Data into JMP

Getting your data into JMP is a familiar process. Like many other desktop applications, you can simply select **File > Open** to import your data into JMP. JMP can handle many different data formats. Table 2.1 shows the default formats JMP recognizes. Other previously installed applications could contain proprietary formats that might also appear as import options. You can import files with these formats as well.

In this section, we show you how to open JMP data tables and how to import Microsoft Excel spreadsheets and text files in JMP. Each of these file formats follow the same basic procedure, but they have special options that allow you to import exactly what you want.

JMP interfaces with databases using Open DataBase Connectivity standard (ODBC). Through the Database Open Table dialog box, you can query your data using SQL. We illustrate only the essential connectivity here; more information about querying your data is available in the JMP documentation (**Help > Search > SQL**).

At the end of this section, we show you how to create a new data table in JMP.

File Type	File Extension
JMP Files	.jmp, .jsl, .jrn, .jrp, .jmpprj, .jmpmenu
JMP Data Tables	.jmp
Excel 97-2003	.xls
Text Files	.txt, .csv, .dat, .tsv
JMP Scripts	.jsl, .txt
JMP Journals	.jrn
JMP Reports	.jrp
JMP Projects	.jmpprj
JMP Menu Archives	.jmpmenu
SAS Data Sets	.sas7bdat, .sd7, .sd2, .sd5, .ssd01, .xpt
SAS Program Files	.sas
HTML Files	.htm, .html
FACS Files	.fcs
Minitab Portable Worksheet Files	.mtp
MS Access Database	.mdb, .accdb
Excel Files	.xlsm, .xlsx, .xlsb
dBASE Files	.dbf, .ndx, .mdx
All Files	

Table 2.1

Data

Opening a JMP File

Let's start by opening a JMP data table. At the top left of the JMP window is the **File** menu:

1. Select **File > Open** (see Figure 2.1). A familiar dialog box opens (the Macintosh file open dialog box has a different look but similar function).

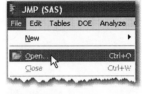

Figure 2.1

2. We will use the Big Class data table described earlier. Click on the **Big Class.jmp** file and select **Open** (see Figure 2.2).

*Note: To locate this file for the first time, select **File > Open > C: > Program Files > SAS > JMP > 8 > Support Files English > Sample Data > Big Class.jmp**. Alternatively, you can also select **Help > Sample Data > Open the Sample Data Directory > Big Class.jmp**.*

Figure 2.2

These steps open the JMP data table Big Class shown here (see Figure 2.3). With these simple steps, you are now ready to analyze or visualize this data.

This spreadsheet-like table is referred to as the *JMP data table*. It appears in JMP in the same format as any data imported from another application. Section 2.2 discusses the components of the data table.

Note: In the Big Class example, the variables (name, age, sex, height, and weight) are located as column heads and the individual sets of observations appear in rows. This structured format of the data is required. The importing examples that follow assume that your data already exists in this format. Section 2.6 introduces some tools to use if your data does not adhere to this structure.

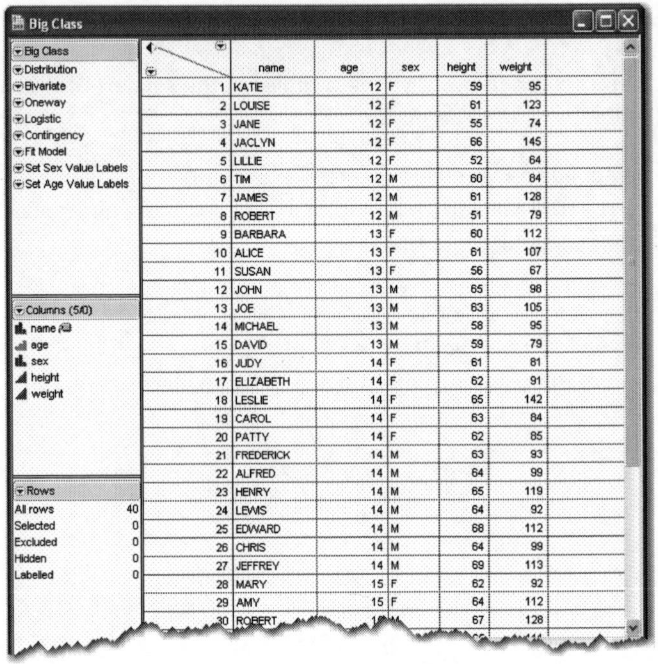

Figure 2.3

Importing Data into JMP

Importing data into JMP from another file format is similar to opening a JMP file. Within the **File** > **Open** pop-up window, the **Files of Type** drop-down menu indicates **ALL JMP Files** as the default.

If you are importing another file type, simply click on the down arrow and select the right type. You can also select **All Files** from the drop-down menu (see Figure 2.4). Select the file that you want, and then click **Open**.

*Note: If you know the format of your data, first select the correct format from the **Files of Type** drop-down menu. You will see the available files of that type within the folder. Once you've located the right file, select the file and click **Open**.*

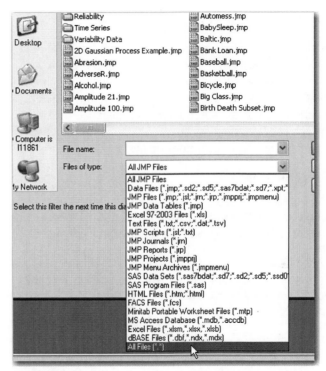

Figure 2.4

Data

Importing an Excel File

Importing an Excel file is easy as long as your variables are in columns and your cases or sets of observations are in rows. Any variable names should appear in the row directly above the first row of data, as shown to the right. (If they are not, see Section 2.6.) By default, the import process automatically opens and converts the data into a JMP data table and uses your variable names as column headings:

1. Select **File** > **Open.**

    The **Big Class.xls** file, which is illustrated here, can be found by selecting **C:** > **Program Files** > **SAS** > **JMP** > **8** > **Support Files English** > **Sample Import Data** > **Big Class. xls.**

2. From the **Files of Type** drop-down menu, select **Excel 97-2003** (or **Excel Files** for files created in Excel 2007 with the .xlsx file extension).

3. Select the file that you want, and then select **Open** (see Figure 2.5).

[Shortcut: If you have JMP open and an Excel worksheet or workbook on your desktop, you can simply drag the file over the JMP shortcut icon on your desktop to import the file.]

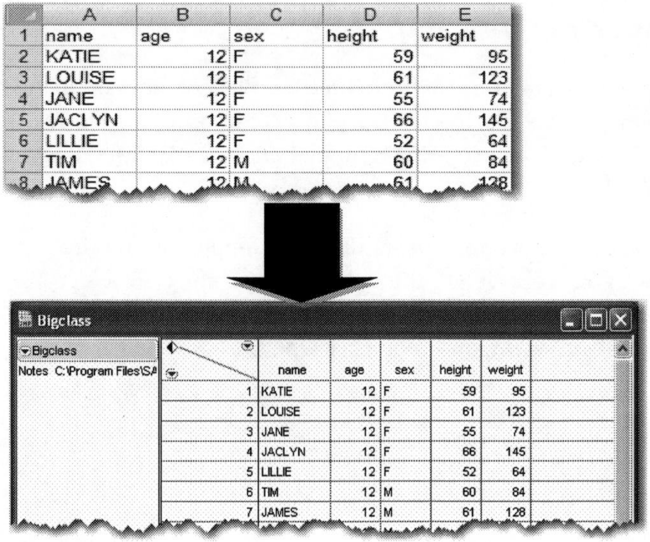

Figure 2.5

Note: If you have multiple worksheets within an Excel workbook, JMP's default import process opens all of them. For these cases, JMP provides additional controls in the Excel Open dialog box to select individual worksheets and to specify that your headings are placed in the right location within the JMP data table.

Importing a Text File

If you are importing a text file, a handy wizard is included in the **Data with Preview** file option. This wizard allows you to view your data and specify how you want it to appear before importing it into a JMP data table. It also provides options to convert your text file if it is delimited by commas, tabs, or spaces:

1. Select **File > Open.**

    The **Big Class.txt** file, which is illustrated here, can be found by selecting **Help > Sample Data > Open the Sample Data Directory > Sample Import Data > Big Class.txt**.

2. Select **Text Files (*.txt, *.csv, *.dat, *.tsv)** in the **Files of Type** drop-down menu. Select the file.

3. Select **Data with preview** (see Figure 2.6).

4. Select **Open.**

5. Choose the settings you need (see Figure 2.7). Click **Next** and then **Import.**

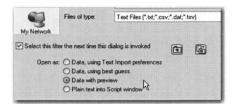

Figure 2.6

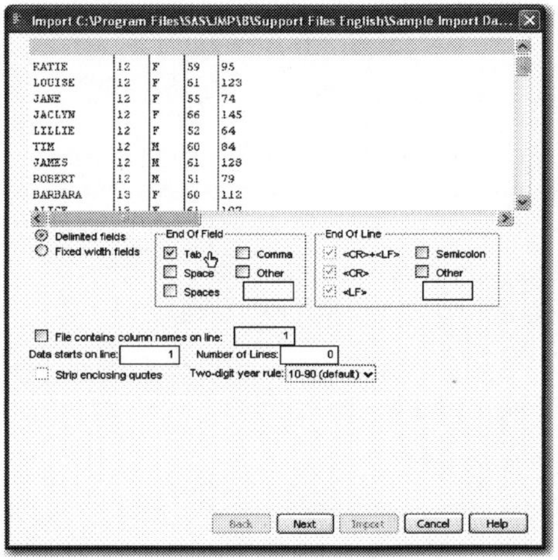

Figure 2.7

Data

Importing a Database File

Options to import data extracted from a database are available through ODBC within JMP. To access this data, first connect to the database (the data source should already be defined) and then specify the table of interest. You can also query your data using the **Advanced** button. If you need more help defining your data source, select **Help > Books > JMP User Guide > Ch. 2 Creating and Opening Files.**

1. Select **File > Database > Open Table** (see Figure 2.8).

2. The Database Open Table window appears (see Figure 2.9). It prompts you to connect to your database and either to open a data table or to specify a query. Clicking the **Connect** button launches the Select Data Source window to locate and connect to your database.

3. Locate the table of interest, highlight it, and click **Open Table** to import the data.

Figure 2.8

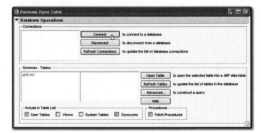

Figure 2.9

*Note: You have the option of directly importing Microsoft Access or dBASE files by selecting **File > Open** as previously discussed (assuming these programs are installed). Using this more direct option allows you to import only a single table.*

Creating a JMP Data Table from Scratch

Select **File > New > Data Table** to create a new data table (see Figure 2.10):

1. Click twice (do not double-click) on the first column's heading and type the column name (the variable name).

2. Press **Enter** and type the data into the first cell directly below the heading. Press **Enter** again, type the data, and repeat as needed. Rows within JMP are consecutively numbered as observations or cases (see Figure 2.11).

3. To create another column, **double-click** on the next column's heading and enter the data as you did before.

If it is more practical for you to enter a series of data for each row as you build your data table, set up all of your column headings first and then use the **Tab** key to move from the left columns to the right. When each column has been filled, the **Tab** key moves down to the beginning of the next row.

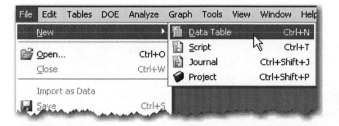

Figure 2.10

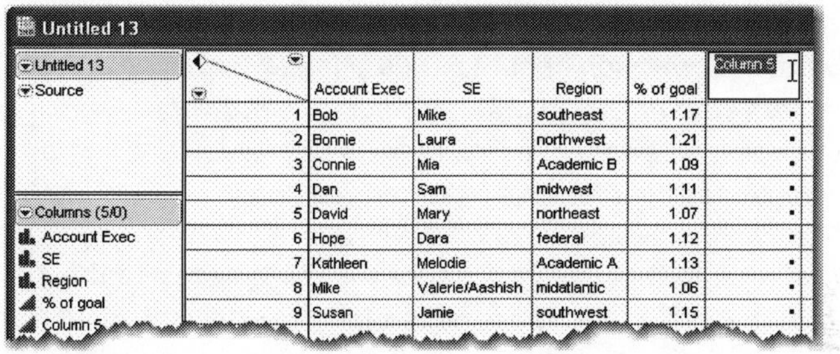

Figure 2.11

Note: JMP will recognize the type of data you are entering and assign a data type to the column, either numeric or character. It also assigns an icon next to the columns (or variables) in the box on the left. These icons are discussed in Section 2.3.

2.2 The JMP Data Table

The JMP data table looks very much like any spreadsheet (see Figure 2.12). In JMP, column headings indicate variables (what you've measured) and rows indicate individual cases or sets of observations. JMP requires your data to be structured this way. If it is not, JMP can help you reformat your data (see Section 2.6).

***Data table** refers to the spreadsheet-like grid where your data resides. The three panels on the left of the data table contain information about your data (metadata). The data grid can contain any number of columns (your variables) or rows (observations or cases). In this sense, we refer to data within the JMP data table as structured data.*

In addition to the data grid, notice the three panels to the left of the data table. These panels provide vital information about your data as well as options to streamline and save your analyses.

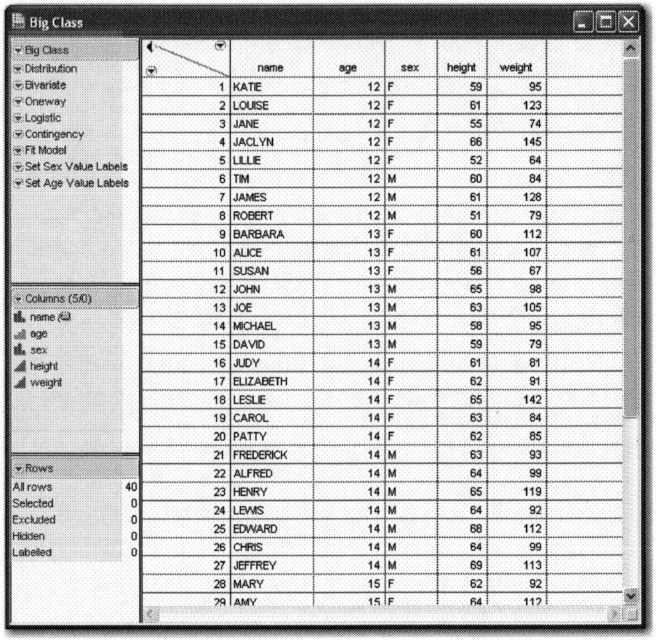

	name	age	sex	height	weight
1	KATIE	12	F	59	95
2	LOUISE	12	F	61	123
3	JANE	12	F	55	74
4	JACLYN	12	F	66	145
5	LILLIE	12	F	52	64
6	TIM	12	M	60	84
7	JAMES	12	M	61	128
8	ROBERT	12	M	51	79
9	BARBARA	13	F	60	112
10	ALICE	13	F	61	107
11	SUSAN	13	F	56	67
12	JOHN	13	M	65	98
13	JOE	13	M	63	105
14	MICHAEL	13	M	58	95
15	DAVID	13	M	59	79
16	JUDY	14	F	61	81
17	ELIZABETH	14	F	62	91
18	LESLIE	14	F	65	142
19	CAROL	14	F	63	84
20	PATTY	14	F	62	85
21	FREDERICK	14	M	63	93
22	ALFRED	14	M	64	99
23	HENRY	14	M	65	119
24	LEWIS	14	M	64	92
25	EDWARD	14	M	68	112
26	CHRIS	14	M	64	99
27	JEFFREY	14	M	69	113
28	MARY	15	F	62	92
29	AMY	15	F	64	112

Figure 2.12

The first and upper-most panel contains the name of the data table (see Figure 2.13). This box stores references and/or scripts. *Scripts* allow you to save, automate, and customize analyses. If you perform a regular analysis or a scheduled task, you will want to learn more about JMP scripts (see the *JMP Scripting Guide* at **Help > Books > JMP Scripting Guide**).

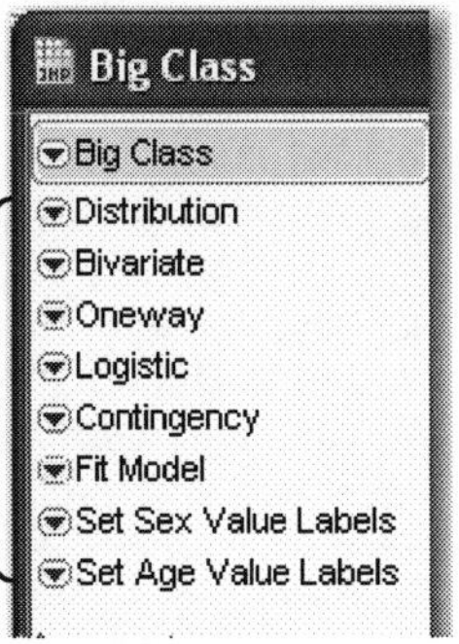

The Columns panel (see Figure 2.14) is where your column names (or variables) appear. Each column has an icon in front of it.

These icons correspond to the modeling type of the data in each column. As discussed in the next section, this is vitally important. JMP produces only the graphs or statistics that are appropriate for a column's modeling type. In most cases, you can change the modeling type by simply clicking on the icon and selecting another appropriate format.

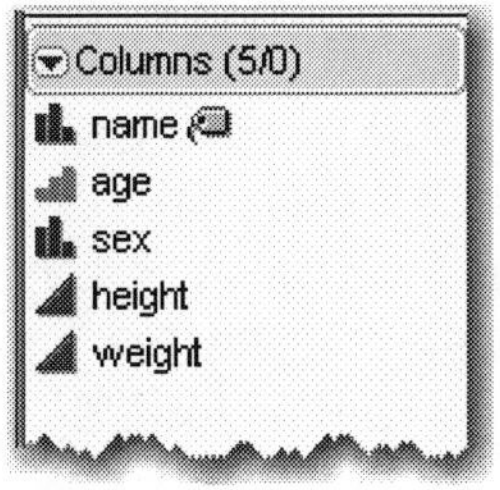

Figure 2.13

Figure 2.14

The bottom panel is the Rows panel (see Figure 2.15). The Rows panel indicates how many rows (sets of observations) are in your data table. This panel also indicates the number of selected, hidden, or excluded rows, if any.

*Note: When rows are hidden, the observations are not included in graphs. When rows are excluded, they are not included in analyses. This state is effective when you want to see or analyze a subset of your data. You can also both hide and exclude specific rows, which effectively deletes the row(s) from your analysis or graph but not from your data table. See Section 6.3 for more information on hiding and excluding rows. (Hiding and excluding does not apply to the **Control Chart** platform.)*

Multiple data tables can be open at any time, but only one active data table can be analyzed at a time.* If you have multiple data tables open within JMP and you want to toggle to a different data table, go to the drop-down menu (see Figure 2.16) and select the data table that you want to use.

Figure 2.15

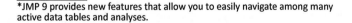

Figure 2.16

Note: There is no practical limit on the size of the data table you can analyze. However, because JMP runs in your computer's local memory, the amount of RAM you have determines the upper size limit of your data table. Your computer should be equipped with 2.5 times the size of the data table. Thus if you have 2 GB of RAM, you can analyze about a 10-variable data set with 1 million rows! Beginning with JMP 7, a JMP Pro version is also available, allowing much larger data files to be analyzed. More details on JMP system requirements can be found at www.jmp.com.

---

*JMP 9 provides new features that allow you to easily navigate among many active data tables and analyses.

2.3 Data and Modeling Types

One of JMP's great features is the ability to produce graphs and statistics that make sense for the data you are analyzing. This feature assumes that your data is correctly classified in the data table. So, what do we mean by data type and modeling type? Let's define a few terms.

Data refers to any values placed in the cell of a JMP data grid. Examples include numeric and/or text descriptions: 3.6, $2500, Female, somewhat likely, or 3/17/10.

Data type refers to the nature of the data. The data type can be either numeric (numbers) or character (often words and letters but sometimes also numbers). The key difference is whether to treat the data as a number (numeric) or a category (character).

Modeling type refers to how the data within a column should be used in an analysis or a graph. JMP uses three distinct modeling types: continuous, nominal, and ordinal.

- *Continuous data* (also referred to as *ratio* and *interval scale data*) takes a numeric form and is often thought of as measurements of some type. For example, home selling prices, income earned, costs per square feet, and dates are all examples of continuous data. As a rule of thumb, continuous data can be calculated. For example, calculating the average cost per square foot would be meaningful.

- *Nominal data* is categorical data (also referred to as *discrete, count,* or *attribute data)* and can take on either a character or numeric form. Nominal data fits into categories or groups such as car type, gender, department, and sales territory and also includes indicator variables like yes/no or 0/1. In nominal data, it is helpful to count the frequency of the occurrence of values, but otherwise nominal data is not used in calculations. For example, calculating the average car type would not be meaningful.

- *Ordinal data* is categorical data that has an inherent order or hierarchy. For instance, Likert scales (such as levels of satisfaction) in a survey and grade levels in school (freshman, sophomore, junior, senior) are examples of ordinal data. That is, they represent categories that have some sequence or order that should be retained in any analysis. Ordinal data is less common than continuous and nominal data, but there are a few analyses designed specifically for it. In most JMP analyses, nominal and ordinal data are treated the same way.

Modeling Type		Data Type	
		Numeric	Character
	Continuous	Yes	No
	Nominal	Yes	Yes
	Ordinal	Yes	Yes

From Table 3.1 Combinations of Data and Modeling Types, *JMP® 8 User Guide, Second Edition.* Go to **Help>Books>JMP User Guide.**

In our example, **Big Class** contains five variables (or columns) representing each of these modeling types (see Figure 2.17). Let's briefly explain why they are classified by their data and modeling types:

Name is nominal because it is a character data type and the student's name is arbitrary.

Age is ordinal because the values are rounded down and we want to retain the six ordered age groups (12 to 17) in our analysis. *Note: Age could also be considered continuous because the values are numeric, but this would treat age differently and yield different results.*

Sex is nominal because its data type is character (M or F) and it has no order.

Height and **weight** are continuous because they are both numeric and represent a measurement.

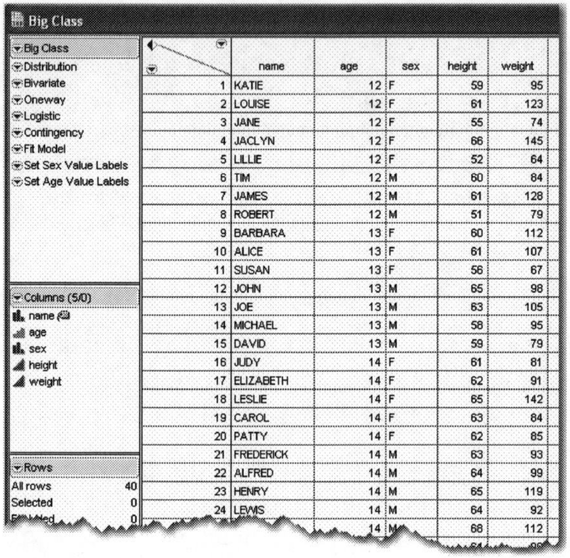

	name	age	sex	height	weight
1	KATIE	12	F	59	95
2	LOUISE	12	F	61	123
3	JANE	12	F	55	74
4	JACLYN	12	F	66	145
5	LILLIE	12	F	52	64
6	TIM	12	M	60	84
7	JAMES	12	M	61	128
8	ROBERT	12	M	51	79
9	BARBARA	13	F	60	112
10	ALICE	13	F	61	107
11	SUSAN	13	F	56	67
12	JOHN	13	M	65	98
13	JOE	13	M	63	105
14	MICHAEL	13	M	58	95
15	DAVID	13	M	59	79
16	JUDY	14	F	61	81
17	ELIZABETH	14	F	62	91
18	LESLIE	14	F	65	142
19	CAROL	14	F	63	84
20	PATTY	14	F	62	85
21	FREDERICK	14	M	63	93
22	ALFRED	14	M	64	99
23	HENRY	14	M	65	119
24	LEWIS	14	M	64	92
		14	M	68	112

Figure 2.17

Note: Row State is a third data type that allows you to store and manage information about a row of data. For more information, select ***Help > Books > JMP User Guide > Ch. 5 Properties and Characteristics of Data > Using Row State Columns.***

Changing the Modeling Type

When you import data, the JMP default selects and assigns one of two modeling types based on whether the data is numeric or character. Numeric data becomes continuous and character data becomes nominal. Sometimes you might want to change the default modeling type of your data to generate results that are more meaningful.

If we imported the **Big Class** data from Excel, for example, age would be imported as a continuous column. We would want to change that to ordinal. Changing the modeling type is simple in JMP. Click on the column's icon in the Columns panel in the data table and select the right type (see Figure 2.18). If the **Continuous** option is grayed out, your data type is classified as character. To change the data type, double-click on the column heading and change the data type to numeric (see Figure 2.19). In this window, you can also change the modeling type along with a host of other formatting options, which are described in the next section.

Figure 2.18

Figure 2.19

For more information, select **Help > Books > JMP Stat and Graph Guide**.

2.4 Formatting Data

Sometimes data isn't in the best shape or in the right form when it is imported. Fortunately, JMP has extensive formatting abilities. This section focuses on the most common features, including:

- Cleaning up your data format, such as decimal places, dates, times, and currency. We will use the Column Info window to accomplish these tasks.

- Introducing the Formula Editor, which allows you to create new columns from old ones, add IF statements, and transform data using basic or more advanced functions. We will introduce a basic example in this section. For more information, select **Help > Books > JMP User Guide.**

- Learning to use the RECODE command, which is a handy way to merge similar categorical responses into a single category. For example, if you have Woman, Female, and Girl as responses, you can merge these into a single response, Female.

Example 2.2 Movies

We will use the **Movies.jmp** data table to illustrate the concepts in this section. This data table consists of the 277 top-grossing movies released between 1937 and 2003. The columns are:

Movie: name of movie

Type: genre/category of movie (for example, comedy, family)

Rating: US movie rating system (for example, general audience [G], adult [R])

Year: year of movie release (for example, 1937)

Domestic $: US domestic revenue in $ earned by the movie in that year

Worldwide $: Worldwide revenue in $ earned by the movie in that year

Director: director of movie

Help > Sample Data > Business and Demographic > Movies

You can access this data table in the **Sample Data** folder that is installed with JMP by selecting **File > Open > C: > Program Files > SAS > JMP > 8 > Support Files English > Sample Data > Movies. jmp.**

Getting your data into a standard format is done through the Column Info window. Options to format your data are driven by the data and modeling types specified for that column of data. You can change these types, if necessary, to meet the requirements of your formatting. Recall that changing these types affects the graphs or statistics you can generate from that column (see the previous section). Let's begin by opening the **Movies.jmp** data table:

1. Open the **Movies.jmp** data table.

2. Select the **Domestic$ column**, and then select **Cols > Column Info.**

3. Because Domestic$ is a numeric value, you see the **Format** drop-down menu (see Figure 2.20), which leads to several options. It is also our starting point for the next items we'll discuss. If you select a Character column, the **Format** menu does not appear in the Column Info window.

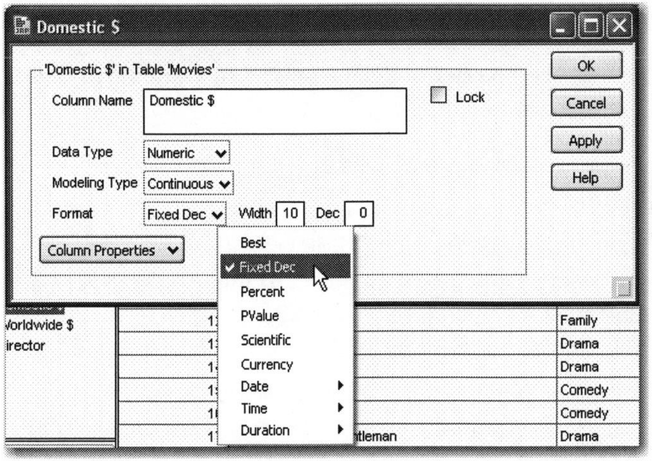

Figure 2.20

*Note: You can also either double-click on the column name as mentioned in the previous section, or right-click on the column and select **Column Info** from the menu.*

Formatting Decimal Places

To change the number of decimal places in a column of data, do the following:

1. Click on the column of interest. In our example, it is **Domestic$**.

2. Select **Cols > Column Info.** JMP will make a best guess on the format of the data; in our example, **Currency** was correctly specified (see Figure 2.21). You can easily change this format by selecting another format from the menu.

3. To the right of the **Format** menu are two boxes, **Width** and **Dec**. **Width** refers to the data table column width, and **Dec** refers to the number of decimals right of the point. In our example, type 0 in the **Dec** box, and then select **Apply** (see Figure 2.22).

*Note: This procedure applies to both **Percent** and **Fixed Dec**.*

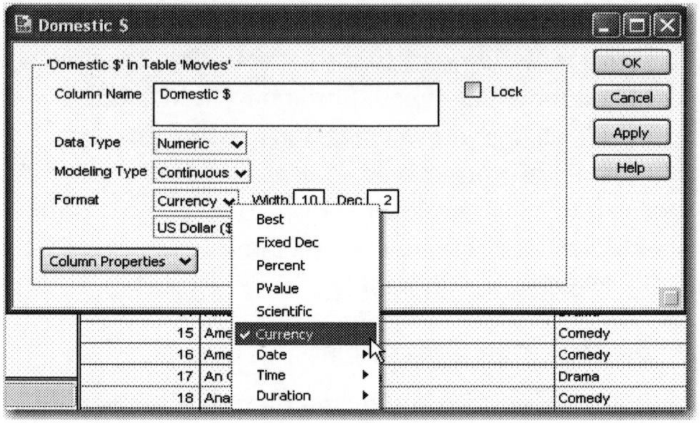

Figure 2.21

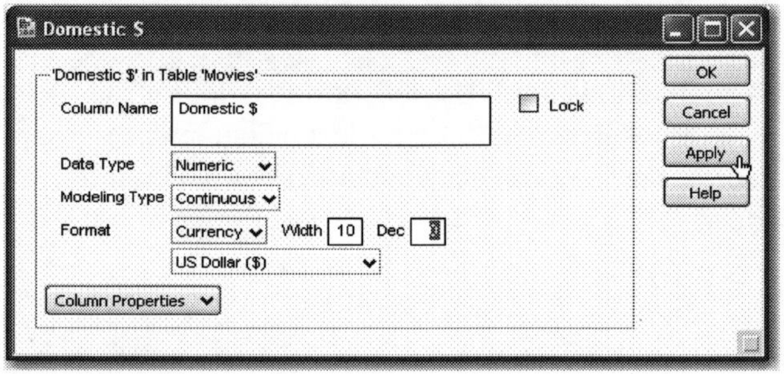

Figure 2.22

Formatting Dates, Time, and Duration

Dates are numeric values in JMP, which allows them to be transformed into other date formats and calculated for duration or elapsed time. If you are importing data that contains dates, ensure that the data type is numeric.

The Column Info (**Cols > Column Info**) window provides the column's information (see Figure 2.23) and several date format options, as seen on the following page. When a date is selected from the **Format** menu, a secondary drop-down menu for the display format appears, along with a similar drop-down menu for the input format of your imported data. The format of your imported data needs to match one of JMP's input format options, which can then be transformed into any format among the display format options.

Let's walk through a new example, TechStock, to illustrate this concept.

Example 2.3 TechStock

We will use the TechStock data table to illustrate dates in this section. This data set contains the stock price of the NASDAQ 100 (QQQ), at the high, low, and close for each trading day during the period 11/27/2000 to 2/26/2001. You can access this data set in the **Sample Data** folder that is installed with JMP:

File > Open > C: > Program Files > SAS > JMP > 8 > Support Files English > Sample Data > Techstock.jmp

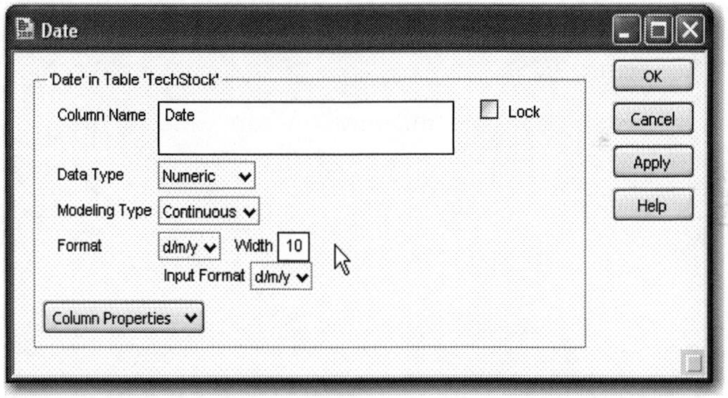

Figure 2.23

Data

1. **Open** the **TechStock.jmp** data table.

   TechStock.jmp can be found at **Help** > **Sample Data** > **Open the Sample Data Directory** > **TechStock.jmp**.

2. Click on the **Date** column name.

3. Select **Cols > Column Info**. Open the **Format** drop-down menu, and select **Date**, which displays how the dates will appear in the data table.

4. It is currently displayed as d/m/y (see Figure 2.24), as indicated by the check mark. Change the format to **Monddyyyy.** Click **Apply** or **OK.** The date is now displayed as abbreviated month, day, and year in the Date column (see Figure 2.25).

You can also format time and duration from this window.

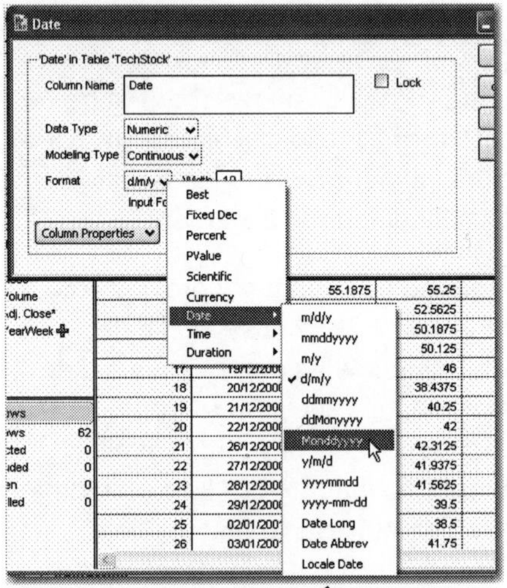

Figure 2.24

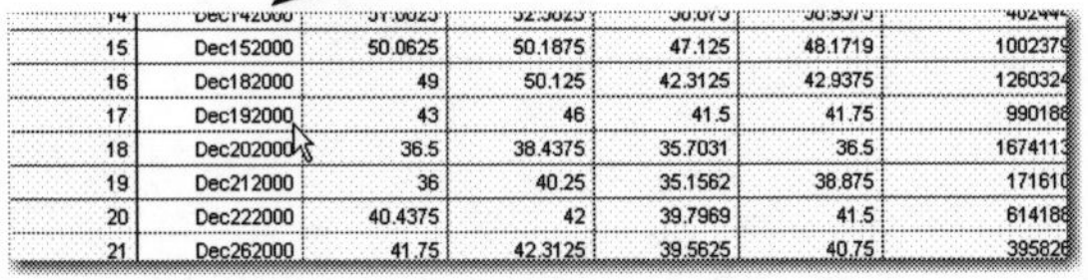

Figure 2.25

Column Properties Menu

Column Properties, another useful tool in the Column Info window (see Figure 2.26), allows you to add formulas, check ranges of values for auditing, and assign customized ordering to the data, among other tasks. These functions are described in detail in the *JMP User Guide* (**Help > Books > JMP User Guide**).

Formula Editor

JMP's formula editor is handy and flexible. Use it when you need to create a new column that contains values that are calculated or derived from existing columns in your data table. You can also transform your data, add conditional statements, and much more. Due to the advanced nature of these features, we'll cover only the most basic features here. For more information, see the *JMP User Guide* at **Help > Books > JMP User Guide**.

Column Properties ▼

Formula
Notes
Range Check
List Check
Value Labels
Value Ordering
Value Colors
Axis
Coding
Mixture
Row Order Levels
Spec Limits
Control Limits
Response Limits
Design Role
Factor Changes
Sigma
Units
Distribution
Time Frequency
Other...

Figure 2.26

Data

Data

One of the common operations performed with the Formula Editor is creating a new column of data that contains a calculation from existing columns. To illustrate this feature, let's return to our **Movies.jmp** data table. For example, say we want to obtain the international revenues from these movies by subtracting the domestic revenues (Domestic$) from the worldwide revenues (Worldwide$).

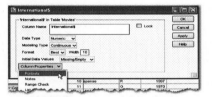

Figure 2.27

1. First, we need to create a new column. Double-click in the column head to the right of our last populated column (Director) (see Figure 2.27). Type **"International$"** in the heading, and press Enter. Click in the column head to highlight.

2. Select **Cols > Column Info**. The Formula window appears in the **Columns Properties** menu (see Figure 2.28). Select **Formula** and **Edit Formula**. You see a list of columns or variables on the left side of the window.

Figure 2.28

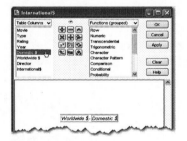

3. Click on the **Worldwide$** column, then the "-" symbol in the center palette, and then the **Domestic$** column. You see your formula take shape in the preview window (see Figure 2.29).

4. When you click **OK**, the calculated values appear in the column of your data table.

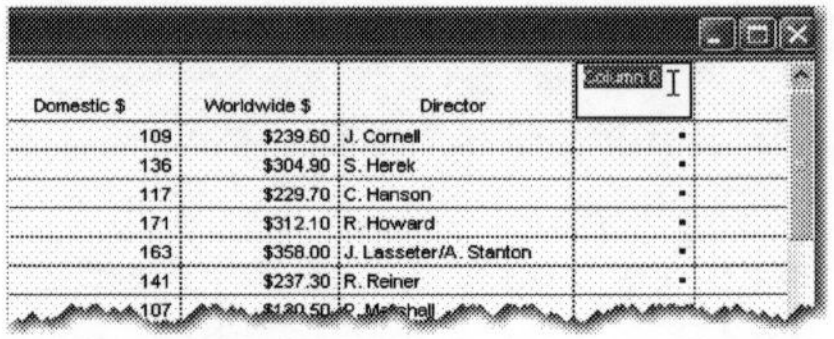

Figure 2.29

Value Ordering

Value ordering allows you to specify an order to the values of a categorical column. JMP's built-in defaults order common ordinal columns such as months of the year or days of the week, but there are other instances when you'd like to arrange responses (values) in some logical order for graphs and analyses. For example, some surveys have a range of responses, from "Not Satisfied" to "Very Satisfied" with a few intermediate responses in between.

In the **Movies.jmp** data table, we want to reorder the rating of the movies so they are displayed in this order: G, PG, PG-13, and R.

1. Click on the **Rating** column. Select **Cols > Column Info**, and then select **Column Properties**.

2. Select **Value Ordering** (see Figure 2.30). A new window appears with the available responses from the Rating column. Select a response and move it up or down or reverse the order, whichever is appropriate. Select **Reverse** for this example (see Figure 2.31).

3. When you are satisfied with the order, click **Apply** and **OK**.

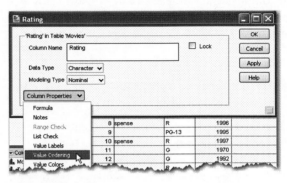

Figure 2.30

Figure 2.31

Now let's see the results of this exercise by employing the distribution platform, which is discussed in Chapters 3 and 5. Here's a preview:

4. Go to **Analyze > Distribution** (see Figure 2.32).

5. In the Distribution window, select **Rating**, **Type**, and **Year**, and click **Y, Columns**. Click **OK** (see Figure 2.33).

6. Three bar graphs appear side by side. Although R was the first response listed in the Value Ordering window, it appears last at the bottom of the Distribution graph shown at right. If we did not make this change, the order of these responses would be reversed.

7. Click on the green **G** bar under Rating (see Figure 2.34). The G responses are now highlighted in the G bar as well as those same responses reflected by Type and Year. This dynamic visual feature is available in all JMP graphs.

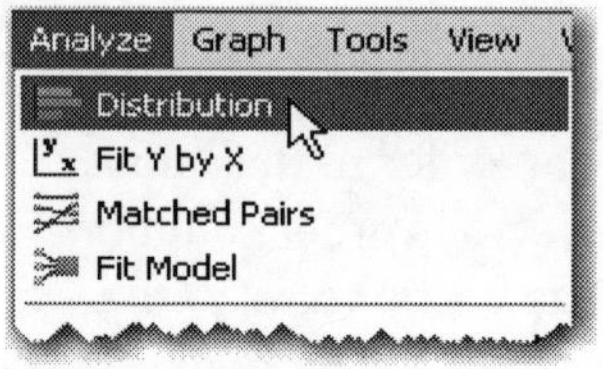

Figure 2.32

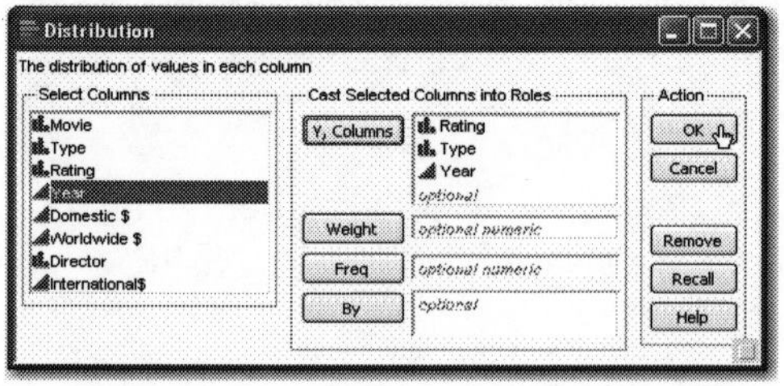

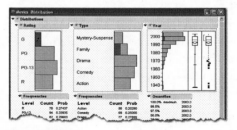

Figure 2.33

Figure 2.34

Recode

The Recode command is useful when you have a column of data that contains values that you'd like to rename or consolidate. For example, if you have data labeled iPod, Nano, iTouch, and Classic, you might want to consolidate these into one response, iPod. Recoding assigns a specified new value to all of the existing responses of the original name or value.

In the **Movies.jmp** data table, we want to replace PG-13 movies with a PG rating because many movies made before 1985 only contained ratings of G, PG, and R.

1. Select the column you want to recode, **Ratings**. Select **Cols > Recode** (see Figure 2.35).

2. This command generates an input window of current and unique responses, with an area to the right to specify a new value. In the box to the right of PG-13, type **PG**, then click **OK** (see Figure 2.36).

Figure 2.35

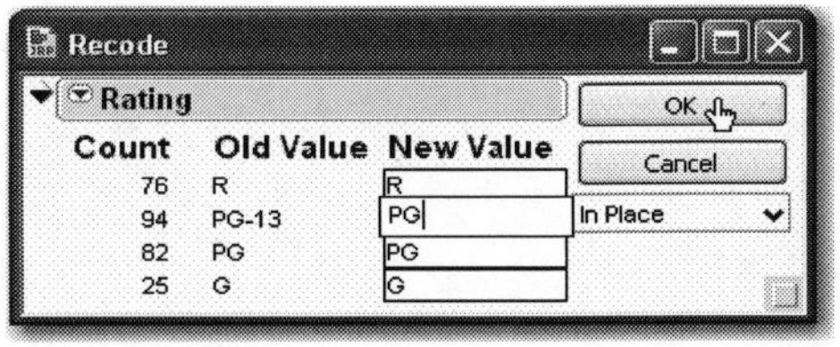

Figure 2.36

Note: Once you've selected a new value, you can replace that value in the same column, create a new column with these values, or even create a formula column. Be careful! If you select **in Place***, these values cannot be changed because the Recode command replaces values in that column.*

2.5 Adding Visual Dimension to Your Data

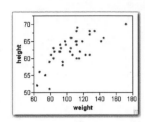

JMP is designed to be visual. Its many useful tools help you visualize or communicate your data effectively. For example, you can use colors or unique markers to signify a range or value of another column in any appropriate graph. Any color or marker assigned to your data can be saved and used in any number of graphs. You can change these colors or markers at any time. We return to our **Big Class.jmp** data table to illustrate this feature:

1. To access the **Rows** menu, select the red triangle* for Rows (see Figure 2.37).

2. Select **Color or Mark by Column.**

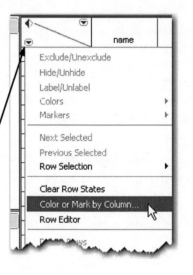

Figure 2.37

---

*Red triangles (or hot spots) appear throughout JMP and provide context-specific options.

3. Select the column you'd like to distinguish with color (sex, in this example). You can see how JMP will express these values in color on the right side of the window (see Figure 2.38). Once you are satisfied, click **OK** and you will see colored markers preceding the row numbers in the data table.

4. Alternatively in the same window, you can also distinguish points by using unique markers (for example, symbols). Like colors, unique markers can be assigned to categorical or continuous columns. Click on the **Markers** drop-down menu (see Figure 2.39). JMP provides many different marker types, and a submenu allows you to view and select the desired type.

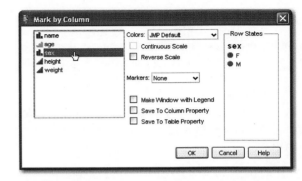

Figure 2.38

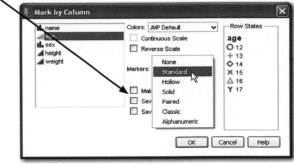

Figure 2.39

Note: You can color by one column and use distinct markers for a second column by first completing this process to apply color on one column and then repeating this process to apply markers on another column.

Adding Labels to Data

Sometimes in the process of exploring your data, it is useful to identify a point by a name, territory, or product type rather than its row number in the data table. Adding labels allows you to see these identifiers in a graph by simply clicking on a point of interest. For example, in the **Big Class.jmp** data table, we want to see the name of the student (rather than a row number) in a graph:

1. First, select a column name by clicking on the column name in the data table. This selection activates the **Label option** when selecting the Columns red triangle/hotspot. Select **Label/Unlabel** (see Figure 2.40).

2. You then see a label or what might look like a price tag next to that column in the Columns panel of the data table (see Figure 2.41). When creating a graph, labels with the name of the student (rather than the row number) are displayed when they are selected with the mouse.

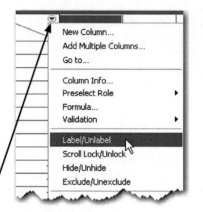

Figure 2.40

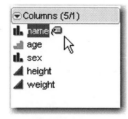

Figure 2.41

2.6 The Tables Menu

Tables are a collection of JMP tools you'll need to use to manage your data, whether you're sorting it, transposing it, or joining multiple data tables. Put another way, if your data is not structured in a manner that fits the JMP analysis framework, you need to use tables to improve the structure. To keep things simple, we'll cover just a few of these features, including sorting, joining, and dealing with missing data. In this section, we learn

- How to structure your imported data into a form JMP will recognize. We use the **Table** menu for these operations.

- What to do when you have missing data.

Using the **Big Class.jmp** data table, let's first take a quick look at the **Summary** option under the **Tables** menu. This command allows you to obtain a variety of summary statistics for any column.

- Select **Tables > Summary** (see Figure 2.42). Choose **height** from Select Columns and **Mean** from the **Statistics** menu (see Figure 2.43). Click **OK**.

Figure 2.42

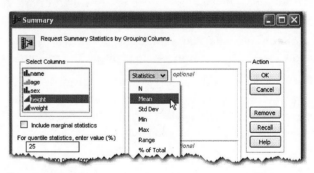

Figure 2.43

Sorting

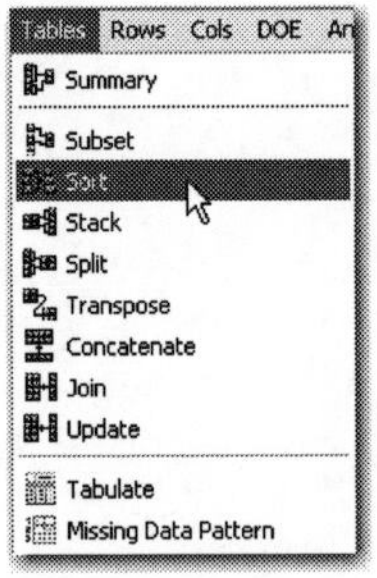

You can sort numeric columns from highest to lowest or lowest to highest. With character columns, you can sort character data by alphabetical or reverse alphabetical order. Using JMP's sorting option keeps the rows (your sets of observations) intact. Sorting also creates a new JMP data table with the sorted values (if you check **Replace table**, the sorted values replace the existing data table).

1. Select **Tables > Sort**. In the resulting window, identify which column you want to sort. Select **height** and click **By.**

2. Click on the column(s) you want to sort in the right window (height, in our example) to highlight the column.

3. Select the way you want to sort them, highest to lowest or lowest to highest, using the corresponding pyramid icon (see Figure 2.44). Click **OK.** Each entire row is sorted according to the conditions you apply.

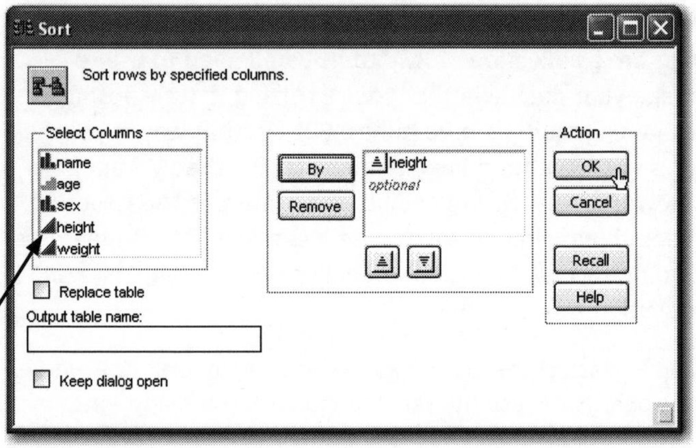

Figure 2.44

More information on sorting is available in **Help > Books > JMP User Guide > Reshaping Data.**

Joining

The **Join** option from the **Tables** menu allows you to combine or merge two or more different data tables into one. If the columns in your original data have the same name and type, this is a simple process. If not, there are some handy JMP tools to help you select how two different data tables can be joined. Let's look at a simple example:

1. First open the data tables you'd like to join. Select **Tables > Join**.

   Trial1.jmp and **Trial2.jmp** can be found at **Help > Sample Data > Open the Sample Data Directory > Trial1.jmp** and repeat for **Trial2.jmp**.

2. The window indicates your active data table and prompts you to select another data table that you want to merge or join. Select the data table(s) you want to join (see Figure 2.45). The column headings of each appear in the Source Columns windows.

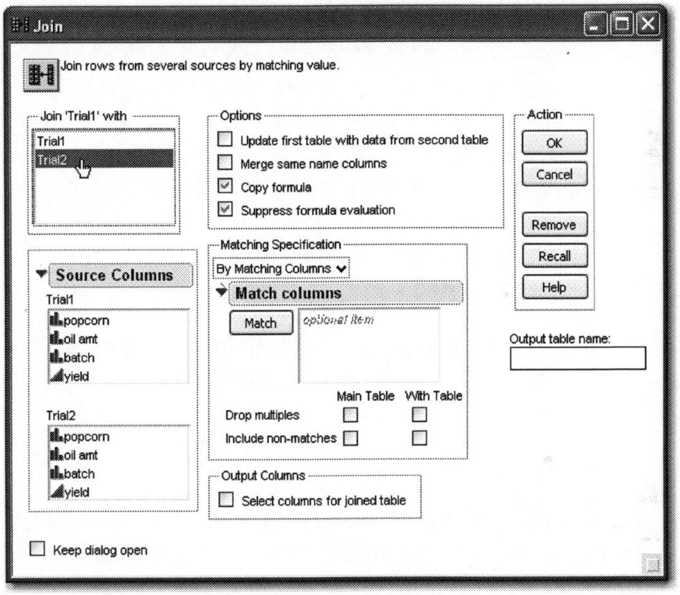

Figure 2.45

3. Decide how you want to join the data under the **Matching Specification** drop-down menu.

   a. If your data has different column headings or you want to select a subset of columns, use **By Matching Columns** (the default in JMP 8). Click on a column from each of the **Source Columns** windows you'd like to match and click **Match**. You now see each of those selected columns in the Match columns window with an "=" symbol between them.

   b. **By Row Number**, the default in JMP 7, joins your data side-by-side by its row number.

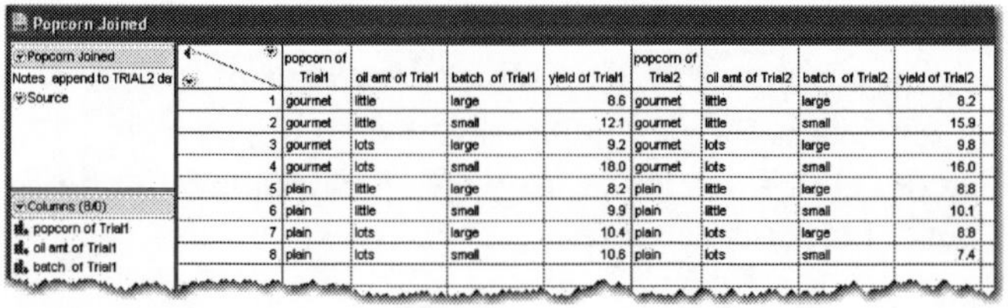

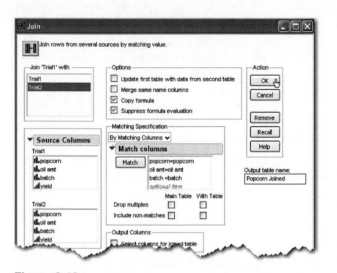

4. If you want to name the new data table, enter the name in the Output table name box (otherwise, it will be named Untitled), and click **OK** (see Figure 2.46). A new data table appears (see Figure 2.47).

Figure 2.46

	popcorn of Trial1	oil amt of Trial1	batch of Trial1	yield of Trial1	popcorn of Trial2	oil amt of Trial2	batch of Trial2	yield of Trial2
1	gourmet	little	large	8.6	gourmet	little	large	8.2
2	gourmet	little	small	12.1	gourmet	little	small	15.9
3	gourmet	lots	large	9.2	gourmet	lots	large	9.8
4	gourmet	lots	small	18.0	gourmet	lots	small	16.0
5	plain	little	large	8.2	plain	little	large	8.8
6	plain	little	small	9.9	plain	little	small	10.1
7	plain	lots	large	10.4	plain	lots	large	8.8
8	plain	lots	small	10.8	plain	lots	small	7.4

Figure 2.47

Missing Data

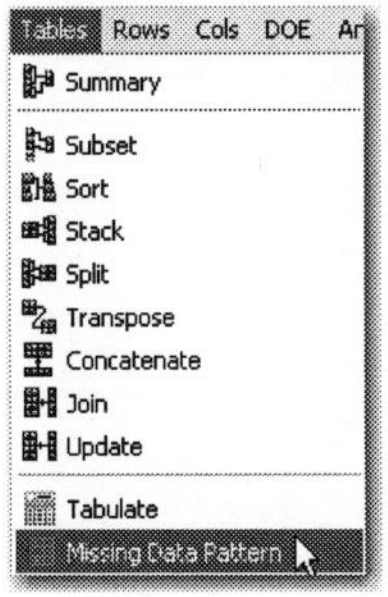

The Missing Data Pattern window can help you identify the quantity of missing data or whether any patterns exist due to non-response, data importing, or data entry errors. The **Missing Data Pattern** feature under the **Tables** menu searches your specified columns and summarizes the frequencies of missing data. To illustrate this feature, some values from the **Big Class.jmp** data table have been removed:

1. Select **Tables > Missing Data Pattern**. This generates the window on the right (see Figure 2.48).
2. Select the columns in the left panel that you want to search. Select **Add Columns**, and then click **OK**.
3. This command generates a new Missing Data Pattern table that contains a count of rows that have missing values and a count of rows that have the same missing values among the same column(s) (see Figure 2.49).

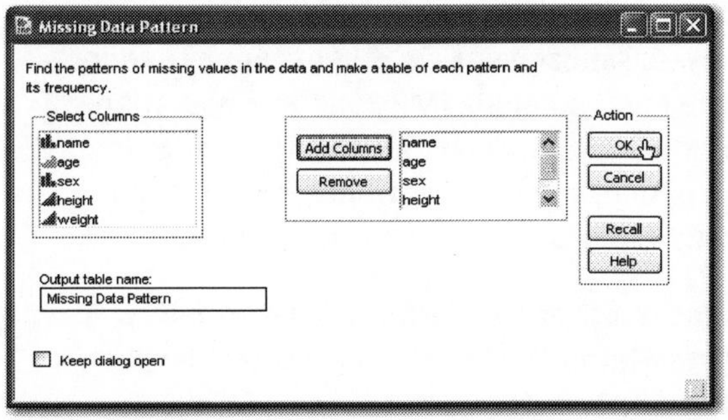

Figure 2.48

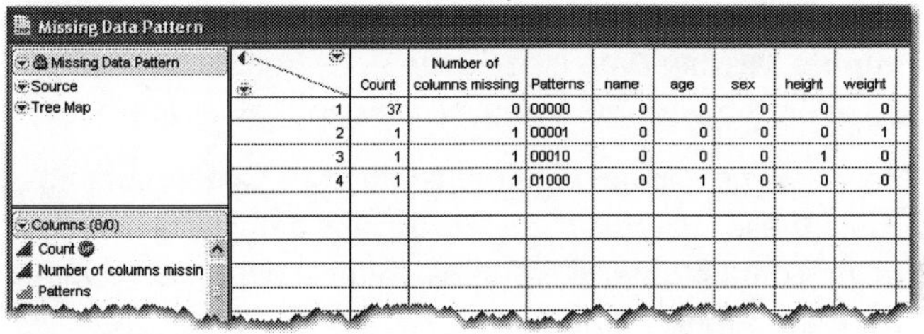

	Count	Number of columns missing	Patterns	name	age	sex	height	weight
1	37	0	00000	0	0	0	0	0
2	1	1	00001	0	0	0	0	1
3	1	1	00010	0	0	0	1	0
4	1	1	01000	0	1	0	0	0

Figure 2.49

*Note: You can proceed without addressing missing values, but JMP will ignore (or exclude) any rows with missing values in the analysis. Exceptions to this rule can be found in the Partition platform, Stepwise and Fit Model with the **fit each separately** option).*

One solution to missing continuous data is to impute them. *Imputing* analyzes similar values in other columns and rows to estimate the missing value. JMP 8 has an imputation feature under the red triangle within the Multivariate window. To illustrate, some values from the height and weight columns in **Big Class.jmp** have been removed:

1. First, run the multivariate platform. Select **Analyze > Multivariate Methods > Multivariate**. Select **height** and **weight** (the continuous columns), in the **Y, Columns** window, and then click **OK** (see Figure 2.50).

2. Click the red triangle next to **Multivariate** and select **Impute Missing Data** (see Figure 2.51). A new data table is generated with the estimated missing values in place.

3. Because you can only impute continuous values, cut and paste these columns into your original data table, which might contain other data types. (Alternatively, use **Update** from the **Tables** menu.)

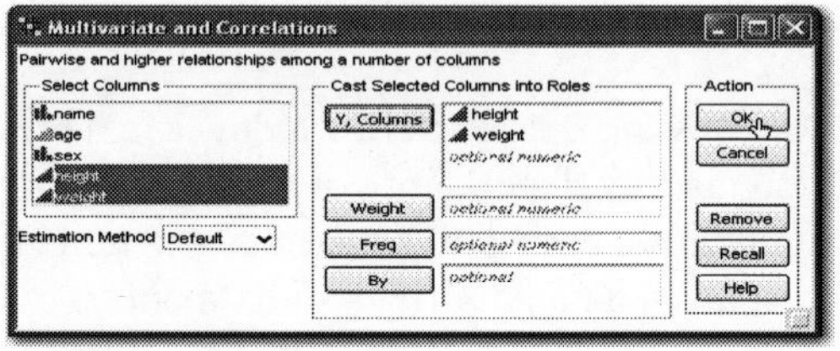

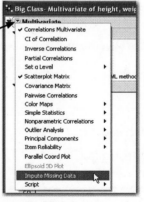

Figure 2.50

Figure 2.51

Note: Methods of handling missing values are best selected with the help of an expert.

2.7 Summary

In this chapter, we covered a wide range of topics on getting your data into JMP and learning how to manage it. Because the data table not only stores your data but also stores key information that drives the appropriate analysis and graphs, it serves as the critical starting point for all exploration and visualization within JMP.

Analyzing and visualizing data often requires special features, and there are many advanced features in JMP that we didn't address. As we've indicated, your copy of JMP includes extensive documentation, which you can access through the **Books** section under the **Help** menu. We recommend the *JMP User Guide* for a complete discussion of data, data tables, and the **Tables** menu. Select **Help > Books > JMP User Guide** (see Figure 2.52).

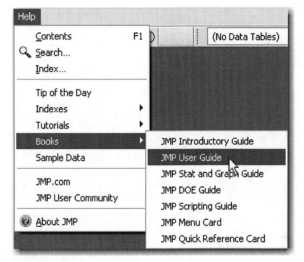

Figure 2.52

Data

Chapter 3 • Index of Graphs

This chapter is a quick reference to commonly used graphs in JMP. The format of this chapter is different from other chapters, for good reason. In this chapter, you can peruse the graphs much like you'd look through a cookbook—find what you want and follow the recipe. A picture of the graph, brief description, required data conditions, usage description, and the steps required to generate the graph immediately follow. This chapter is not intended to be a complete index of graphs available in JMP, but we have tried to pick out those that we see used most often. This chapter is for the user who knows what graph they want and can pick it out by how it looks or what it's called.

You will find that many graph windows have additional options that allow you to further enhance your graphical result. However, we have focused on the steps to generate the base case of each graph, which are illustrated in the figures that accompany each graph.

Some of the graphs illustrated in this chapter are accessed from the **Graphs** menu, while others are accessed from the **Analyze** menu. Statistical output is provided with graphs generated from the **Analyze** menu. For instructions on sharing or printing graphs and on pasting graphs into other applications such as Microsoft Word or PowerPoint, see Section 7.5.

You can customize the appearance of any graph (including colors, markers, axes, legends, and fonts) by simply right-clicking on the area or item you want to change. See Section 2.5 and Section 7.2 for more details on customizing the appearance of your graphs.

Graphs

3.1 Basic Charts

The Chart platform produces hundreds of different charts for general purposes. The platform responds differently depending on the data types and the roles of the columns you select in the Chart window. This section is designed to help you understand some of the commonly needed charts that can be produced with this multi-purpose platform.

Figure 3.1

The Chart platform is accessed from the Graph menu (see Figure 3.1). It produces graphs for every numeric column specified. Some character columns also produce a chart.

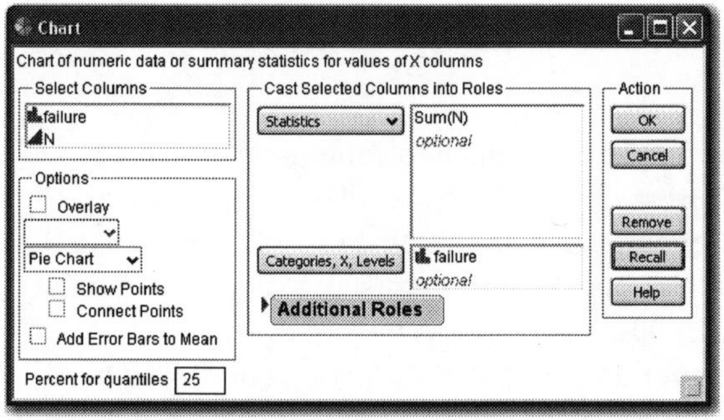

The **Categories**, X, **Levels** role (Figure 3.2) is always treated as categorical. This role is optional for some charts and required for others.

If column attributes like numeric, character, and categorical are unfamiliar to you, see Section 2.3.

By default, a vertical bar chart appears, but there are options to show horizontal bar charts, line charts, needle charts, point charts, or pie charts as we'll see in this section.

Charts are useful for making graphical representations of summary statistics.

Figure 3.2

Graphs

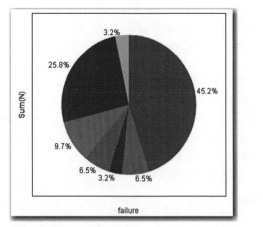

Figure 3.3

Pie Chart: A circular chart divided into areas proportional to the percentages of the whole or total.

Usage: Used when representing proportions, percentages, or fractions of any measured quantity. Some examples are market share, customer preferences, and percents of any kind of category or group. As shown, pie charts can display percentages of seven different failure types (see Figure 3.3).

Required: One nominal column and a continuous column to define the size of the pie chart slices.

Other column combinations are supported. See **Help > Search**. Type **Pie Chart** in the search field.

Select **Graph > Chart**. Then select **Pie Chart** from the **Options** drop-down menu. Select a continuous column and then from the **Statistics** drop-down menu, select a summary such as **Sum**. Select a nominal column, move it to the **Categories, X, Levels** role, and click **OK**. To add percentages to the pie slices as depicted, select the red triangle, and then select **Label Options > Label by Percent of Total Values**.

Data table used for pie chart example: **Help > Sample Data > Control Charts > Failure**.

Graphs

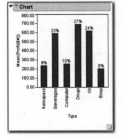

Figure 3.4a (Bar Chart)

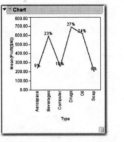

Figure 3.4b (Line Chart)

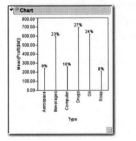

Figure 3.4c (Needle Chart)

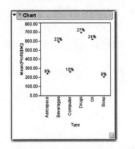

Figure 3.4d (Point Chart)

Bar Chart, Line Chart, Needle Chart, and Point Chart are all related: Charts with bars, lines, needles, or points showing lengths that are proportional to quantities; as visualized here, showing percentages.

Usage: Similar to pie charts representing proportions, percentages, or fractions of any measured quantity. Some examples are market share, proportions, and customer preferences expressed as percents of any kind of category or group. As shown, these charts display average profits for six company types in four chart styles (see Figures 3.4a, 3.4b, 3.4c, and 3.4d).

Required: One continuous, nominal, or ordinal column for the **Statistics** role. **Optional:** A continuous, nominal, or ordinal column for the **Categories, X, Levels** role.

Select **Graph > Chart**. Then select **Bar Chart, Line Chart, Needle Chart**, or **Point Chart** from the **Options** drop-down menu for the chart listed. Select a continuous column and then from the **Statistics** drop-down menu, select a summary such as **Sum** or **Mean**. Optionally select a nominal or ordinal column for the **Categories, X, Levels** role. To add percentages as shown, select the red triangle, and then select **Label Options > Label by Percent of Total Values**.

Data table used for these examples: **Help > Sample Data > Business and Demographic > Financial.**

Graphs

3.2 Control and Variability Charts, Pareto and Overlay Plots

Control Charts

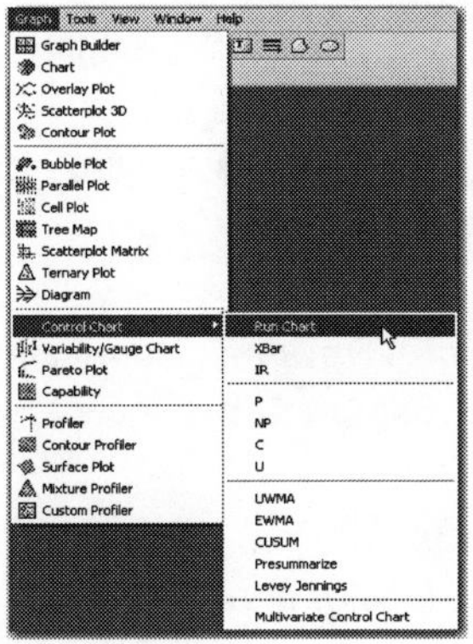

Figure 3.5

This section groups typical graphs used to measure product or service quality. These graphs are common in quality improvement scenarios.

Control charts are graphical and analytic tools for deciding whether a process is in a state of statistical control and for monitoring an in-control process.

Control charts help determine if variations in measurements of a product are caused by small, normal variations that cannot be controlled or by some larger, special cause that can be. The type of chart to use is based on the nature of the data.

Control charts are broadly classified into control charts for measurements and control charts for attributes. Control charts for measurements and control charts for attributes come in several varieties with single or multi-letter or single-letter names attached to them. Control charts appear under the **Graph** menu and contain a submenu of specific control chart types (see Figure 3.5). This section covers some of the more common ones that are included in JMP.

Pareto plots are often included as a quality metric for processes and products. The Pareto Plot command produces charts to display the relative frequency or severity of problems in a quality-related process or operation. The steps to produce a Pareto plot are included in this section.

A *variability chart* plots the mean for each level of a factor, with all plots side by side. Along with the data, you can view the mean, range, and standard deviation* of the data in each category, seeing how they change across the categories. The analysis options assume that the primary interest is in how the mean and variance change across the categories.

Use overlay plots when you want to display individual data points rather than summaries of data points, as we saw in the Chart platform earlier. When these individual data points need to be categorized and subcategorized, overlay plots are useful. An example of a multiple overlay appears in this section.

---

*See Appendix B for a description of statistical terms.

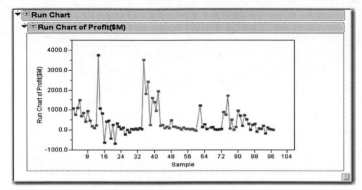

Figure 3.6

Run Chart: Displays a column of data as a connected series of points (see Figure 3.6).

Usage: Displays data in a column. Frequently used as a first visualization of quality data and to assess ranges of variability. Examples include delay times, counts of rejected items, or measurement variability in a product or service. As shown, the run chart displays company profits over a 97-month period.

Required: One or more numeric columns. **Optional:** Nominal, ordinal, or continuous columns for the **Sample Label** and **By** roles.

If data in a column is sorted by time, then the x-axis will display in the time-sorted order. If a row color is assigned as shown, row colors will color the points.

Select **Graph > Control Chart > Run Chart**.

Data table used for this example: **Help > Sample Data > Business and Demographic > Financial**.

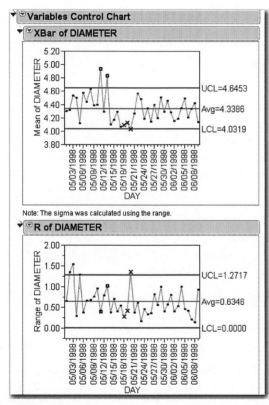

Note: The sigma was calculated using the range.

Figure 3.7

X-Bar/R-Charts: Displays quality characteristics measured on a continuous scale. A typical analysis shows both the process mean and its variability, aligned above a corresponding R-chart or a range or standard deviation chart, respectively.

Usage: Normally used for numerical data that is recorded in subgroups in some logical manner (for example, three production parts measured every hour). A special cause, such as a broken tool, will then appear as an abnormal pattern of points on the chart. As shown, the chart displays several part diameters outside of the control limits (see Figure 3.7).

Required: One or more numeric columns for the **Process** role. **Optional:** Nominal, ordinal, or continuous columns for the **Sample Label** and **By** roles.

Select **Graph > Control Chart > XBar.**

Data table used for these examples: **Help > Sample Data > Control Charts > Diameter**.

Graphs

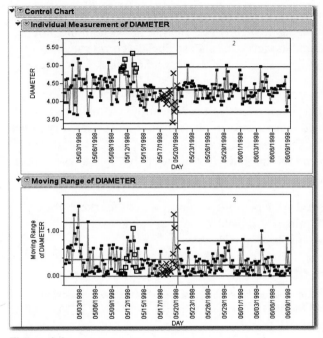

Figure 3.8

IR Chart (with phases): Displays individual measurements. Individual measurement charts are appropriate when only one measurement is available for each subgroup sample.

The accompanying moving range chart displays moving ranges of two or more successive measurements. Moving ranges are computed as the number of consecutive measurements entered in the Range Span box. The default range span is 2.

Usage: Used when only one measurement is available for each subgroup. An example would include samples of acid in pickle vats. IR charts are efficient at detecting relatively large shifts in the process average. As shown, the charts display the contrasting control limits and process performance before and after a quality improvement as phases 1 and 2 (see Figure 3.8).

Required: One or more numeric columns for the **Process** role. **Optional:** Continuous, nominal, or ordinal columns for the **Sample Label**, **Phase**, and **By** roles.

Select **Graph > Control Chart > IR**.

To add phases as shown, select a nominal column for the **Phase** role.

Data table used for these examples: **Help > Sample Data > Control Charts > Diameter**.

Graphs

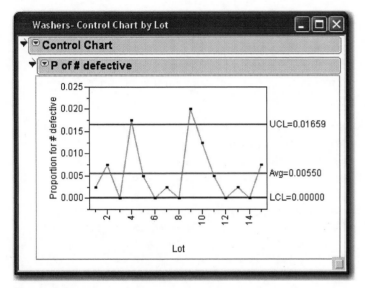

Figure 3.9

P Chart: An attribute chart that displays the proportion of nonconforming (defective) items in subgroup samples (lots), which can vary in size. Because each subgroup for a P chart consists of N items, and an item is judged as either conforming or nonconforming, the maximum number of nonconforming items in a subgroup is N.

Usage: Used when one or more errors might propagate within the same sample (for example, loan transactions where more than one error may occur or where one or more errors may be present in the same unit such as on the surface of a DVD). As shown, the chart displays the proportion of defective washers across many lots of washers within and outside the control limits (see Figure 3.9).

Required: One or more numeric columns for the **Process** role. **Optional:** Continuous, nominal, or ordinal columns for the **Sample Label**, **Phase**, and **By** roles. Sample Size must be a numeric column. A constant or variable sample size can be specified and must be numeric.

Select **Graph > Control Chart > P**.

Data table used for this example: **Help > Sample Data > Control Charts > Washers**.

Graphs

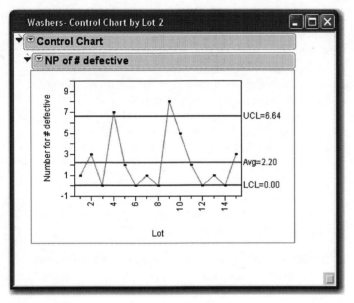

Figure 3.10

NP Chart: An attribute chart that displays the number of nonconforming (defective) items in fixed-sized subgroup samples. Because each subgroup for an NP chart consists of N_i items, and an item is judged as either conforming or nonconforming, the maximum number of nonconforming items in subgroup i is N_i.

Usage: A fixed sample is taken from an established number of transactions or manufactured items each month. From this sample, the number of transactions or items that had one or more errors is counted. The control chart then tracks the number of errors per group or lot. As shown, the chart displays the number of defective washers across many lots of washers within and outside the control limits (see Figure 3.10).

Required: One or more numeric columns for the **Process** role. **Optional:** Continuous, nominal, or ordinal columns for the **Sample Label**, **Phase**, and **By** roles. **Sample Size** must be a numeric column. A constant or variable sample size can be specified and must be numeric.

Select **Graph > Control Chart > NP**.

Data table used for these examples: **Help > Sample Data > Control Charts > Washers**.

Graphs

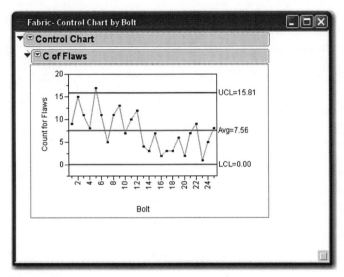

Figure 3.11

C Chart: An attribute chart that displays the number of nonconformities (defects) in a subgroup.

Usage: The values of C computed from each subgroup are plotted on the vertical axis and can then be used to control the quality of the subgroup. As shown, the chart displays the number of flaws in each bolt of fabric within and outside the control limits (see Figure 3.11).

Required: One or more numeric columns for the **Process** role. **Optional:** Continuous, nominal, or ordinal columns for the **Sample Label**, **Phase**, and **By** roles. **Sample Size** must be a numeric column. A constant or variable sample size can be specified and must be numeric.

Select **Graph > Control Chart > C**.

Data table used for these examples: **Help > Sample Data > Control Charts > Fabric.**

Graphs

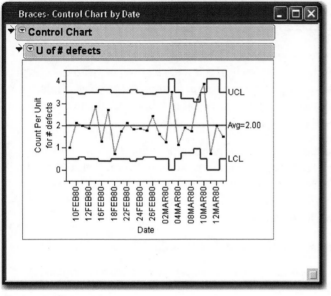

Figure 3.12

U Chart: An attribute chart that displays the number of nonconformities (defects) per unit in subgroup samples that can have a varying number of inspection units.

Usage: To count the number of defective units in subgroups of varying numbers. As shown, the chart counts the number of defective braces in groups of braces of varying size on specific dates and indicates if the count is within or outside of the control limit (see Figure 3.12).

Required: One or more numeric columns for the **Process** role. **Optional:** Continuous, nominal, or ordinal columns for the **Sample Label**, **Phase**, and **By** roles.

Sample Size must be a numeric column.

Select **Graph > Control Chart > U**.

Data table used for this example: **Help > Sample Data > Control Charts > Braces**.

Graphs

Pareto Plot

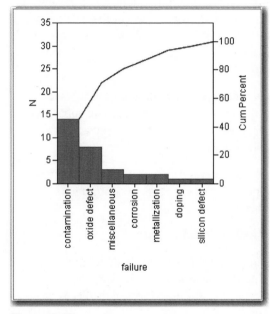

Figure 3.13

Pareto Plot: A chart often included as a quality metric for processes and products. The Pareto Plot produces charts to display the relative frequency of problems in a quality-related process or operation. A *Pareto plot* is a bar chart that displays the classification of problem occurrences, arranged in decreasing order. The column with values that are the cause of a problem is displayed as X in the plot and is called the Cause column. An optional column with values assigning the frequencies is assigned as Freq. An optional column whose value holds a weighting value is assigned as Weight.

Usage: For counts of defects by occurrence of defect causes. Plot can be used to target improvement efforts toward those failures that are most serious or common. As shown, the chart displays defect counts and cumulative percents of seven types of semiconductor defects (see Figure 3.13).

Required: At least one column for the **Cause** role. Additional options for **Frequency** and **Weight** roles.

Select **Help > Search > Pareto Plot** for more information.

To generate the plot, select **Graph > Pareto Plot**.

Data table used for this example: **Help > Sample Data > Control Charts > Failure**.

Variability Chart

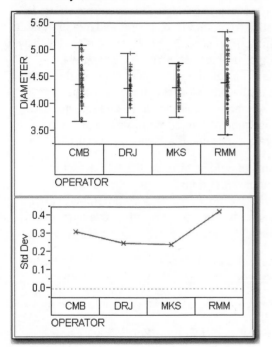

Figure 3.14

Variability Chart: A chart that illustrates how numeric values vary across categories. Along with the data, you can view the mean, range, and standard deviation of the data in each category. The analysis options assume that the primary interest is in how the mean, range, and variance change across the categories.

Usage: For viewing the ranges, standard deviation, and means of a measured column across groups and subgroups. As shown, part diameter variability is displayed across different operators (see Figure 3.14).

Required: At least one numeric column for the **Y, Response** role and at least one nominal, ordinal, or continuous column for the **X, Grouping** role.

Select **Graph > Variability/Gauge Chart**.

Select one column for the **Y, Response** role and one column for the **X, Grouping** role. An optional column for the **X, Grouping** role produces horizontally nested results by the subgroup, overlaid.

Data table used for this example: **Help > Sample Data > Control Charts > Diameter**.

Overlay Plot

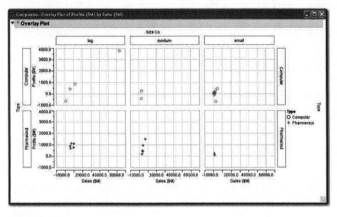

Figure 3.15

Like the variability chart just presented, the overlay plot allows you to visualize values over any specified times or groups. A key difference, however, is that it allows you to specify multiple Y columns and group those values in a meaningful way. An example of an overlay plot with more than one Y value is shown for reference here (see Figure 3.15).

Many of the overlay plot options are also duplicated by graphs available in Graph Builder. See Chapter 4 for more information. A couple of examples of overlay plots follow.

To generate an overlay plot, select **Graph > Overlay Plot** (see Figure 3.16).

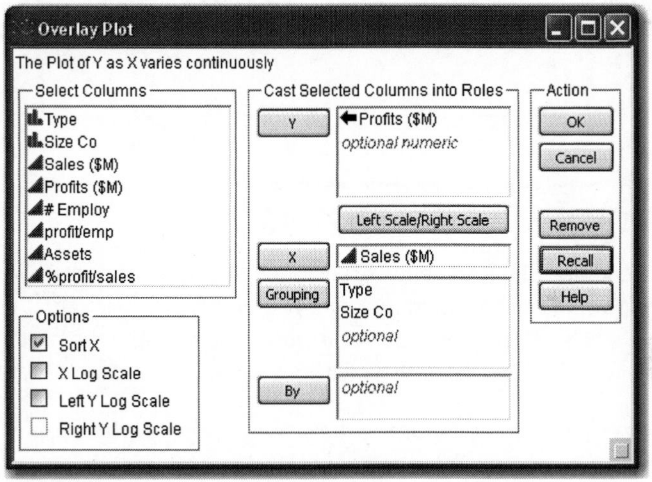

Figure 3.16

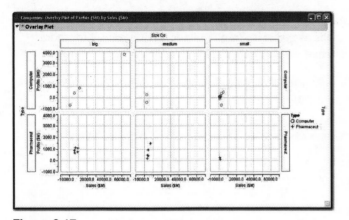

Figure 3.17

The Overlay Plot gives an overlay of a numeric or categorical x column for all specified numeric y columns. The axis can have either a linear or a log scale. You can also show the plots for each y separately, with or without a common x-axis.

Usage: When you want more than one y-axis and when you want nested groups of measurements. For example, the profits of two types of companies are measured against sales, as shown. These are then broken out by company size,

contrasting assets, and profits by color in each company type and size (see Figure 3.17).

Required: One or more numeric columns for the **Y** role and optional continuous, nominal, or ordinal columns for the **X** and **Grouping** roles.

Select **Graph > Overlay Plot**. Select one or more numeric columns for the **Y** role. For an overlay of groups, select a nominal, ordinal, or continuous column for the **Grouping** role. For a labeled horizontal axis, select a column for the **X** role.

Data table used for this example: **Help > Sample Data > Business and Demographic > Companies**.

Note: Many overlay plots are also available using the Graph Builder platform in JMP. See Chapter 4 to explore this method to produce overlay plots.

Graphs

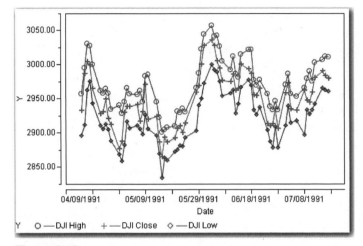

Figure 3.18

Alternate Overlay Plot (Overlay Groups): The overlay plot produces overlays of columns or groups on a single bivariate plot (see Figure 3.18).

Usage: When you want to overlay multiple groups of data into one graph when there is a single x and a single y axis. The chart shows the Dow Jones High, Low, and Close index for a period of time with contrasting colors for High, Low, and Close.

Required: At least two numeric columns for the **Y** role.
Optional: Continuous, nominal, or ordinal columns for the **X** and **Grouping** roles.

Select **Graph > Overlay Plot**. Select at least two numeric columns for the **Y** role, and at least one column for the **X**, **Grouping** role.

Data table used for this example: **Help > Sample Data > Business and Demographic > XYZ Stock Averages (Plots).**

Note: When the overlay chart appears, additional customizations are available on the red triangle.

Graphs

3.3 Graphs of One Column

Unlike the charts introduced so far, graphs in this section are accompanied by statistical results. These graphs depict the distribution of values for one column of data and provide appropriate tools to assess their properties.

These graphs help you understand the nature of a column, such as how widely the values vary or whether there are any curious qualities to the data.

Most of these graphs are found within the Distribution platform from the **Analyze** menu. We also briefly cover time series in this section, as this is often a one-column type graph.

Note: You can choose more than one column with these graphs, but each column will be graphed and analyzed independently side-by-side. When you are looking at more than one column, know that the graphs are linked, which allows you to click on any part of the graph to see and explore those values represented in the graphs of other selected columns. See Chapter 2 for more information on how graphs and data are linked.

Graphs

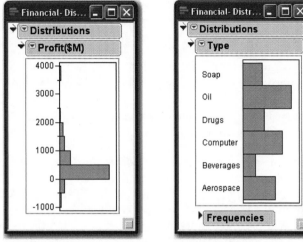

Figure 3.19a

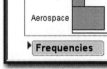

Figure 3.19b

Distribution: Examines properties of a single continuous, nominal, or ordinal column.

Continuous Usage: To view the properties of a continuous distribution such as shape, range, and data density. As shown in Figure 3.19a, the chart displays the profits (or losses) of a selection of technology companies from the late 1990s.

Continuous Distribution Requires: One or more continuous columns for the **Y, Columns** role.

Nominal, Ordinal Usage: Similar to a bar chart, allows you to view the properties of a frequency distribution such as the relative counts or percentages of fixed groups. As shown in Figure 3.19b, the chart displays the frequency of company types as bars for a selection of technology companies from the late 1990s.

Frequency Distribution Requires: One or more nominal or ordinal columns for the **Y, Columns** role.

Select **Analyze > Distribution**. Select a column and place it in the **Y, Columns** role, and click **OK**.

Data table used for this example: **Help > Sample Data > Business and Demographic > Financial.**

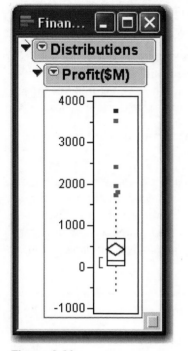

Figure 3.20

Outlier Box Plot: A chart for detecting extreme values and properties of a distribution; sometimes called a *Tukey Box Plot.**

Usage: To view the properties of a continuous distribution such as quartiles, moments, and outliers.* As shown, the plot displays a few very profitable companies as colored points that are well beyond the main body of companies (50% of which are contained in the box) (see Figure 3.20).

Required: One or more continuous columns for the **Y, Columns** role.

Select **Analyze > Distribution**. Select a continuous column and place it in the **Y, Columns** role, and click **OK**. Then click on the red triangle, and select **Outlier Box Plot**. (Program preferences might generate the plot automatically when running the Distribution platform.)

Data table used for this example: **Help > Sample Data > Business and Demographic > Financial**.

---

*See Appendix B for a description of these terms.

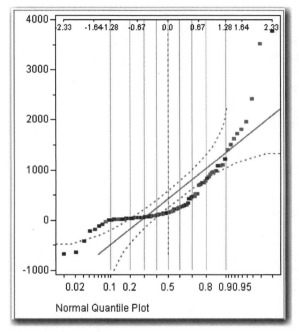

Figure 3.21

Normal Quantile Plot: A chart for visualizing the extent to which a column is normally distributed (that is, bell-shaped). In a symmetrical bell-shaped distribution, the points would fall upon the solid red line in the display and not beyond the confidence curves.

Usage: To view the properties and visually assess the extent to which the data is normally distributed. In this example, the chart displays the profits from a sample of companies and are not normally distributed because most of the observations are not along the solid red diagonal line and some are even beyond the dotted red confidence bands (see Figure 3.21).

Required: One or more continuous columns for the **Y, Columns** role.

Select **Analyze > Distribution >** drag a continuous column into the **Y, Columns** role, and click **OK**. Click on the red triangle, and select **Normal Quantile Plot**.

Data table used for this example: **Help > Sample Data > Business and Demographic > Financial**.

Graphs

Figure 3.22

Mosaic Plot (One Column): A stacked bar chart where each segment is proportional to its group's frequency count.

Usage: To view the properties of a nominal or ordinal distribution, or to visually assess the proportions of data that fall within each group. As shown, the chart displays the proportions or counts of each type of company from a stock portfolio (see Figure 3.22).

Required: One or more nominal or ordinal columns for the **Y, Columns** role.

Select **Analyze > Distribution >** drag a nominal or ordinal column into the **Y, Columns** role, and click **OK**. Click on the red triangle, and select **Mosaic Plot**.

Data table used for this example: **Help > Sample Data > Business and Demographic > Financial**.

Note: Program preferences might generate the plot automatically when running the Distribution platform with a nominal or ordinal column.

Time Series

Time Series is a separate platform that generates a graph of a numeric value over time. It also serves as a platform to employ forecasting techniques and produces statistical results. For more information on these techniques, see the *JMP Statistics and Graphics Guide* (**Help > Books > JMP Stat and Graph Guide**). The Time Series platform is available from the **Analyze** menu and the **Modeling** submenu (see Figure 3.23).

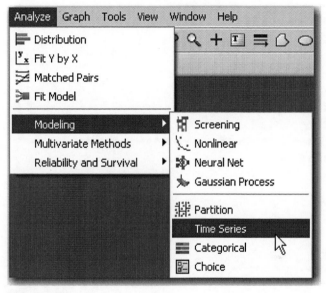

Figure 3.23

Graphs

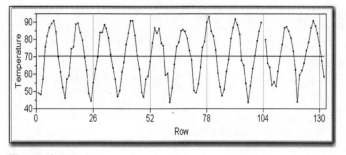

Figure 3.24

Time Series Plot: A graph of numeric values, Y, over a time order, X.

Usage: To view and fit the variability and potential seasonality of a measured value over time. For example, the chart displays the average monthly temperatures, with a clear seasonal trend, in Raleigh, North Carolina, over a 130-month period (see Figure 3.24).

Required: One numeric column for the **Y, Time Series** role.

Options include a numeric Time column (X, Time ID) with corresponding values and an input column. If an X, Time ID column is not specified, JMP orders the data over the rows sequentially.

Select **Analyze > Modeling > Time Series**, drag a continuous measured column to the **Y, Time Series** role, and click **OK**. Select a numeric column for the **Time, ID role**.

Data table used for this example: **Help > Sample Data > Time Series > Raleigh Temps.**

Graphs

3.4 Graphs Comparing Two Columns

The Fit Y by X command studies the relationship of two columns. This command is available from the **Analyze** menu and shows graphs with statistical results for each pair of x and y columns. The type of graph generated by JMP is determined by the modeling types (continuous, nominal, or ordinal) of the columns that are cast into the **X** and **Y** roles. JMP always creates the right graphs based on the modeling type for any function from the **Analyze** menu. In important ways, Fit Y by X is four graphs and analyses in one!

The matrix contained in the Fit Y by X window (see Figure 3.25) provides a visual preview of the graphs that will be generated depending on the modeling type of your Y (the vertical axis) and your X (the horizontal axis).

Note: If the column modeling types nominal, ordinal, and continuous are unfamiliar to you, see Section 2.3.

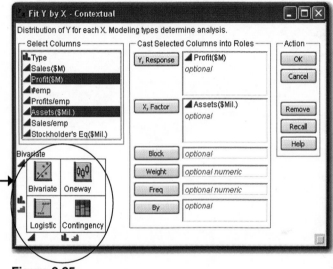

Figure 3.25

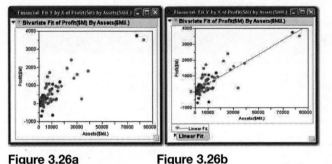

Figure 3.26a   **Figure 3.26b**

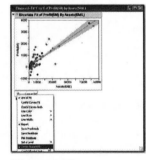

Figure 3.26c

Scatterplot: A graph of the continuous-by-continuous personality within the Fit Y by X command. The analysis begins as a scatterplot of points, to which you can

interactively add a linear fit and confidence curves.

Usage: To view the relationship of a continuous column to another continuous column. An example might be graphing the relationship of profits to assets for a selection of Fortune 500 companies and then fitting a regression line with 95% confidence curves, as shown (see Figures 3.26a, 3.26b, and 3.26c).

Required: One continuous column for the **Y, Response** role and one continuous column for the **X, Factor** role.

Select **Analyze > Fit Y by X,** select a continuous **Y, Response** column and a continuous **X, Factor** column, and click **OK** (see Figure 3.26a).

To add the simple linear least squares fit: From the red triangle next to **Bivariate Fit,** select **Fit Line** (see Figure 3.26b).

To add the confidence shaded curves to the fit: From the red triangle on the **Linear Fit** item, select **Confid Shaded Fit** (see Figure 3.26c).

Data table used for this example: **Help > Sample Data > Business and Demographic > Financial.**

Note: These graphs show colored markers. For more information on how to color or mark rows, see Section 2.5.

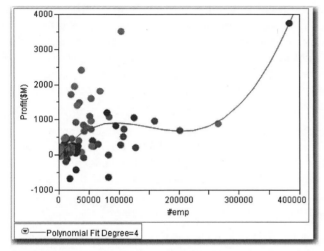

Figure 3.27

Polynomial Fit Degree=4

Scatterplot (with Polynomial fit): A graph that fits a polynomial curve of the degree you select from the **Fit Polynomial** submenu. After you select the polynomial degree, the curve is fit to the data points using least squares regression.

Usage: To view the relationship of a continuous column to another continuous column using a linear polynomial fit where curves produce the best fit of the data. The chart displays the fourth-order polynomial fit showing the relationship of profits to number of employees for a selection of companies (see Figure 3.27).

Required: One continuous column for the **Y, Response** column and one continuous column for the **X, Factor** role.

Select **Analyze > Fit Y by X,** select a continuous **Y, Response** column and a continuous **X, Factor** column, and click **OK**.

From the red triangle, select **Fit Polynomial**, and from the submenu, select a degree number.

Data table used for this example: **Help > Sample Data > Business and Demographic > Financial**.

Graphs

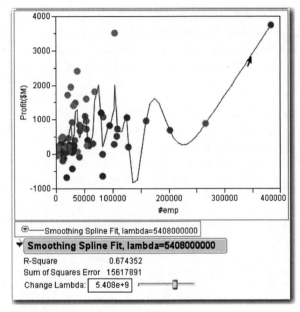

Figure 3.28

Graphs

Scatterplot (with Spline fit): A chart that fits a smoothing spline that varies in smoothness (or flexibility) according to a tuning parameter in the spline formula. Splines contrast the polynomial fit using least squares regression. You can use a spline of varying smoothness to highlight the overall trends in the data without using a linear function to describe the relationship.

Usage: To view the relationship of a continuous column to another continuous column. For example, the chart illustrates that limited profit variation is present in companies with lower numbers of employees. The plot also shows that fewer companies have higher numbers of employees and higher profit variation (see Figure 3.28).

Required: One continuous column for the **Y, Response** role and one continuous column for the **X, Factor** role.

Select **Analyze > Fit Y by X**. Select a continuous column and place it in the **Y, Response** role. Select a continuous column and place it in the **X, Factor** role, and click **OK**.

From the red triangle, select **Fit Spline**, and from the submenu, select the degree of flexibility you want in the spline fit by changing the lambda value.

Data table used for this example: **Help > Sample Data > Business and Demographic > Financial**.

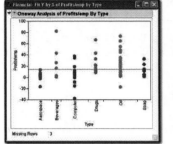

Figure 3.29a

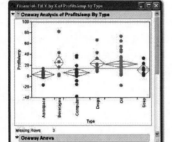

Figure 3.29b

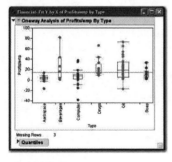

Figure 3.29c

Oneway / Oneway ANOVA / Student's *t* / Quantiles: The Oneway platform analyzes how the distribution of a continuous Y column differs across groups defined by a categorical X column. Group means can be calculated and tested, as well as other statistics and tests. The Oneway platform is the continuous (placed as Y) by nominal/ordinal (placed as X) personality of the Fit Y by X command.

Usage: To view the relationship of a continuous column across the groups (nominal/ordinal) in another column. For example, the chart displays the difference in means and variation in profits and employees across six company types (see Figures 3.29a, 3.29b, and 3.29c).

Required: One continuous column for the **Y, Response** column and one nominal or ordinal column for the **X, Factor** role.

Select **Analyze > Fit Y by X**, select a continuous column for the **Y, Response** role and a nominal or ordinal column for the **X, Factor** role, and click **OK** (see Figure 3.29a).

From the red triangle, select **Means/Anova** (see Figure 3.29b).

From the red triangle, select **Display Options > Quantiles** (see Figure 3.29c).

Data table used for this example: **Help > Sample Data > Business and Demographic > Financial**.

Graphs

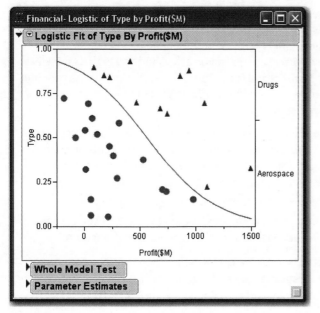

Figure 3.30

Logistic Fit: A chart that estimates the probability of choosing one of the Y response levels as a smooth function of the X factor. The fitted probabilities must be between 0 and 1 and must sum to 1 across the response levels for a given factor value.

In a logistic probability plot, the y-axis represents probability.

Usage: To predict a group or groups by some continuous column. The chart displays a prediction separation (as a probability/percentage) of the type of company (Drugs or Aerospace) by its profits (see Figure 3.30).

Required: One nominal or ordinal column for the **Y, Response** column and one continuous column for the **X, Factor** role.

Select **Analyze > Fit Y by X**, select a nominal or ordinal column for the **Y, Response** role and a continuous column for the **X, Factor** role, and click **OK**.

Data table used for this example: **Help > Sample Data > Business and Demographic > Financial.**

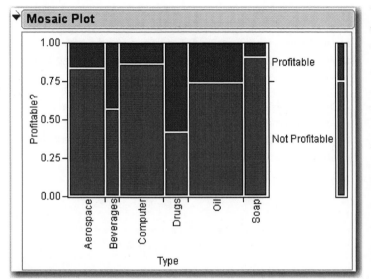

Figure 3.31

Mosaic Plot (two columns): A chart that is divided into small rectangles such that the area of each rectangle is proportional to a frequency count of interest.

The Mosaic Plot appears in the Contingency Platform and is the personality of the Fit Y by X command when both the Y and X columns are nominal or ordinal. Mosaic examines the distribution of a categorical Y column by the values of a categorical X column.

Usage: Group-by-group counts are shown as proportional colored rectangles in a two-by-two arrangement. As shown, the graph displays a simple color chart of the proportion of companies that have been unprofitable or profitable by type. Companies that are profitable are blue and companies that are not profitable are red (see Figure 3.31).

Required One nominal or ordinal column for the **Y, Response** column and at least one nominal or ordinal column for the **X, Factor** role.

Select **Analyze > Fit Y by X**. Select a nominal or ordinal column for the **Y, Response** role and a nominal or ordinal column for the **X, Factor** role, and click **OK**.

Data table used in this example: **Help > Sample Data > Financial**.

Note: The column that denotes "Profitable" and "Not profitable" is produced by an exercise in Chapter 6.

Graphs

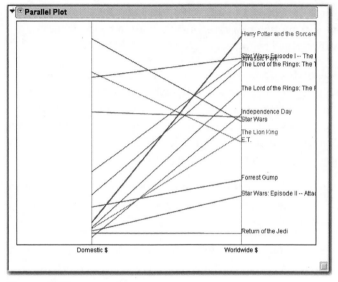

Figure 3.32

Parallel Plot: A plot that displays lines representing differences among specified row values. The strength of a parallel plot is that major differences show up easily as crossed lines. The columns often represent a related measured quantity across time or location.

Usage: To visually detect measurement differences across many rows of data. The chart displays which movies performed better in worldwide revenues relative to domestic revenues and vice-versa (see Figure 3.32).

Required: At least one column of any type; two are recommended.

Select **Graph > Parallel Plot**. Select at least two columns, drag them to the **Y, Response** role, and click **OK.**

To produce the row labels, add a label attribute to the Movie column, select the line in the plot and select **row label**. See Section 2.5 for details on row labels.

Data table used for this example: **Help > Sample Data > Business and Demographic > Movies.**

3.5 Graphs Displaying Multiple Columns

Sometimes it is valuable to see a problem in more than two dimensions. This section uses JMP to visualize three or more columns at once. With the exception of the profiler, which concludes this section, these graphs appear under the **Graph** menu and contain only a few built-in analytic procedures (see Figure 3.33).

Like all JMP graphs, these multi-dimensional (or multi-column) graphs are interactive and allow you to select, rotate, and animate them. You can copy and paste these into other documents and two of them, the bubble plot and the profiler, contain the ability to create exportable Flash files that retain their interactivity and animation.

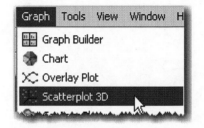

Figure 3.33

Graphs

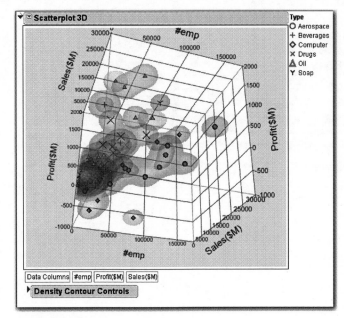

Figure 3.34

Scatterplot 3D: From the **Graph** menu, a chart that displays a three-dimensional scatterplot that can be rotated with your mouse. The Scatterplot 3D platform displays three columns at a time from the columns you select.

Usage: To view patterns among any two or three columns of data. This plot is very useful for exploring data in three dimensions. The chart displays sales (y-axis) by number of employees (x-axis) by profits (z-axis) with colored 3D density ellipses by company type (see Figure 3.34). The eye can detect possible differences among company types across the three columns. This is an interactive plot and can be rotated on any axis. To rotate the graph, click and hold the graph and move the mouse.

Required: Two or more columns of any modeling type (can be continuous, nominal, or ordinal). Three columns are required for a three-dimensional plot.

Select **Graph > Scatterplot 3D**. Select at least two columns (three are recommended) and place them in the **Y, Columns** role, and click **OK**. To include surfaces as displayed, from the red triangle, select **Nonpar Density Contour**.

Data table used for this example: **Help > Sample Data > Business and Demographic > Financial**.

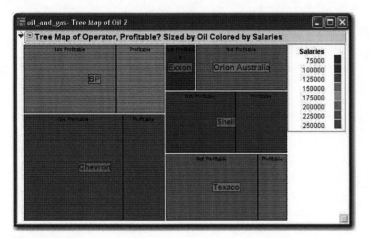

Figure 3.37

Tree Map: A graphical technique of observing patterns among groups that have many levels. Tree maps are especially useful in cases where histograms are ineffective.

Usage: For example, the chart displays oil production by amount, where hot colors represent higher salaries and cold colors represent lower salaries by company. Larger squares are higher oil production amounts as well (see Figure 3.37). These maps produce convenient visual rankings or groups within groups.

Required: At least one continuous, nominal, or ordinal column for the **Categories** role.

Select **Graph > Tree Map**.

Select one nominal or ordinal column for the **Categories** role, and click **OK**. Optionally, select numeric columns and place them in the **Sizes** and **Ordering** roles. Also, optionally select a column and place it in the **Coloring** role, and click **OK**.

To generate a legend as shown, a Coloring column must be set. Then from the red triangle, select **Legend**.

The data table for this example is fictional.

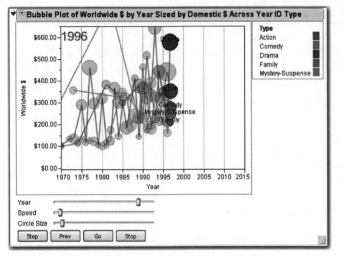

Figure 3.35

Bubble Plot: An interactive scatterplot, which represents its points as circles (bubbles). Optionally the bubbles can be sized according to another column, colored by yet another column, and dynamically indexed by a time column. With the opportunity to see up to five dimensions at once (x position, y position, size, color, and time), bubble plots can produce dramatic visualizations and are effective at communicating complex relationships.

Usage: Summarizing multi-column data in an interactive two-dimensional display. Frequently used where time is one of the columns. The bubble plot in this example displays worldwide revenue from popular movies over the last 90 years by movie type. Sizes of bubbles show domestic US revenue (see Figure 3.35).

Required: One **X** column and one **Y** column of any type (continuous, nominal, or ordinal). For time animation, select a column for the **Time** role.

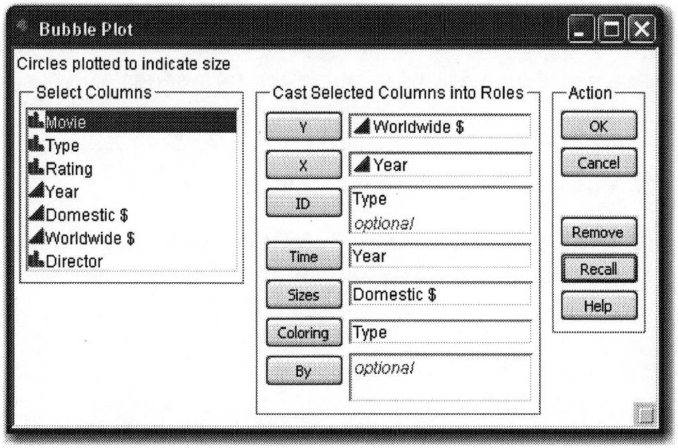

Figure 3.36

Select **Graph > Bubble Plot**.

Select one column and place it in the **Y** role. Select one column and place it in the **X** role. For an animated time series plot, specify a **Time** column. The **ID** column produces a label for each bubble. The **Sizes** and **Coloring** columns can be specified to increase information density (see Figure 3.36).

To generate trail lines (as shown), from the red triangle, select the **Trail Lines** and **Trail Bubbles** options.

To save a Flash file of this plot for viewing in a Web browser or PowerPoint.* From the red triangle, select **Save As Flash (SWF)**, and specify a location for the Flash file.

To turn on the legend, select the red triangle, and select **Legend**.

Data table used for this example: **Help > Sample Data > Business and Demographic > Movies**.

---

*See Section 7.5 for information about placing an animated Flash file into PowerPoint.

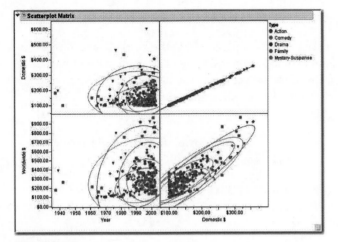

Figure 3.38

Scatterplot Matrix: A chart that provides quick and orderly production of many bivariate graphs. These are assembled so that comparisons among many columns can be conducted visually so that correlation and data pattern can be easily detected. In addition, the plots can be customized and employ advanced features (such as density ellipses) to provide for further analysis.

Usage: In this example, the scatterplot matrix provides every bivariate combination of three columns—domestic revenue, worldwide revenue, and year—for a series of popular movies (see Figure 3.38). This chart quickly produces many correlation plots of all variables specified for easy identification of interesting groups and patterns.

Required: Two or more columns of any type (continuous, nominal, or ordinal) for the **Y, Columns** role. More than two columns are recommended.

Select **Graph > Scatterplot Matrix**.

Select at least one column for the **Y, Columns** role. Select multiple Y and X columns for a matrix of graphs. Optionally, select a column and place it in the **X** role. Select a nominal or an ordinal column and place it in the **Group** role.

To include grouped ellipses as shown, include a column for the **Group** role in the window. Then from the red triangle for the plot, select **Density Ellipses**.

Data table used for this example: **Help > Sample Data > Business and Demographic > Movies.**

Graphs

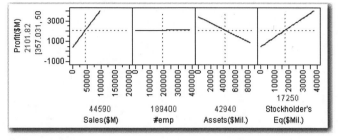

Figure 3.39

Profiler: An interactive graph that provides a simple way to view complex relationships within a model. It lets you visualize what-if scenarios quickly and easily by allowing you to see the effect that changes in one column have on the remaining columns. This tool is especially useful when describing multiple variable models by demonstrating the sensitivity of changes in one or more X columns on the predicted Y.

The profiler displays profile traces for each X column. A profile trace is the predicted response as one column is changed (by dragging the vertical red dotted line in the graphs) while the others are held constant at the current values. The profiler recomputes the predicted responses (in real time) as you vary the value of an X column.

Usage: In this example, the chart displays the relationship between a continuous column Profits($M) and four predictors: number of employees (#emp), sales [Sales($M)], assets [Assets($Mil.)], and stockholder's equity [Stockholder's Eq($Mil.)] (see Figure 3.39). The profiler shows that as sales increase, predicted profits increase, and as assets increase, profits decrease. To generate interactive, real-time predictions for profits, drag the vertical red trace lines in the profiler in JMP (or in the saved Flash file).

Required: A formula column. Can be created in one of two ways: from Fit Model (which generates a formula) or from a formula column entered in a data table by hand.

To create a formula in a column manually, see Section 2.4.

Profiler Creation from Fit Model

Select **Analyze > Fit Model**, and select one or more columns for the **Y** role and one or more columns for the **Construct Model Effects**. Select the **Emphasis** pop-down menu, select **Effect Screening,** and click **Run Model**. The profiler appears at the bottom of the report.

Graphs

Create a Profiler for Use Outside of JMP (Flash File)

From the Fit Model report window from the previous step, select the red triangle, and select **Save Columns > Prediction Formula**. This action saves a prediction formula to a JMP data table column.

Once you have a prediction formula in a column, from the top menu, select **Graph > Profiler**. Select the prediction formula column that has appeared in the data table, place it in the **Y, Prediction Formula role**, and click **OK**. This implementation allows you to save the profiler as a Flash file by selecting **Save as Flash (swf)** from the red triangle in the profiler report.

Data table used for this example: **Help > Sample Data > Business and Demographic > Financial**.

3.6 Summary

This chapter presented a series of the most frequently used graphs and their step-by-step recipes. They are presented in cookbook style so that each can be recognized by its picture, name, or definition and easily replicated with simple steps. These charts represent the first tier of commonly used graphs. The *JMP Statistics and Graphics Guide, Second Edition,* which is included with JMP provides additional graphs you might find useful.

Graphs

Graphs

Chapter 4 • Finding the Right Graph or Summary

4.1 Using Graph Builder
4.2 Using Tabulate
4.3 Summary

In the last chapter, we presented an overview of the commonly used graphs in JMP. This works well as a recipe book when you know what you want your graph to look like. However, at other times, and particularly when you are exploring your data for the first time, it makes sense to take a different approach for several important reasons. First and foremost, you probably don't know what your data is going to tell you. Second, you can learn a great deal about your data when you explore it both visually and numerically. Reviewing your data in both ways can lead you to find the best way to display information.

In this chapter, we take advantage of a key feature of JMP: its ability to create graphs and summaries interactively and intuitively. While all platforms in JMP promote exploration, two features in particular—Graph Builder and Tabulate—are especially designed for this purpose. Because the former is geared at interactively creating graphs and the latter at numerical summaries, we cover them separately. Let us begin with finding the right picture of your data.

Explore

4.1 Using Graph Builder

Graph Builder is a great tool to quickly see your data expressed in many different forms. In many ways, Graph Builder is like a blank canvas waiting for your artistic direction. With Graph Builder, it is helpful to begin to think about your problem and the columns that are central to your questions at hand. Graph Builder is found under the **Graphs** menu when you select **Graphs > Graph Builder** (see Figure 4.2).

Graph Builder works in a drag-and-drop manner. Simply click and hold a column in the **Select Columns** box and drag it (without letting go) to one of the zones located around the canvas. You can hold the column in a zone (without letting go) to get a preview of the graph and then move it to another zone to see an alternative display. Once you release the mouse button, Graph Builder keeps that selection and allows you to tailor the display or select another column to begin exploring relationships. You can repeat this process by adding more columns to view more complex displays.

Example 4.1: TechStock

We'll be working with one of the sample data files to illustrate the steps in this section. The file we'll use, **Techstock.jmp**, contains closing price data from selected technology stocks taken from a period in the early 2000s. The columns include:

Date: the date expressed as dd/mm/yyyy

Open: the opening price of the stock

High: the highest price of the stock on that date

Low: the lowest price of the stock on that date

Close: the closing price of the stock on that date

Volume: the number of shares traded on that date

Adj. Close: the adjusted price of the closing price

YearWeek: the week of the year expressed as yyyy/week#

You can find the data set at **Help > Sample Data > Business and Demographic > Techstock.**

---

1. First open the sample data table as indicated previously. In the sample data table Techstock, the Volume column modeling type is nominal. It makes sense for the next exercise to change the Volume columns from **Nominal** to **Continuous**.
2. In the data table, change the **Volume** column from **Nominal** to **Continuous** by clicking on the icon in the Columns panel on the left (as in Figure 4.1).

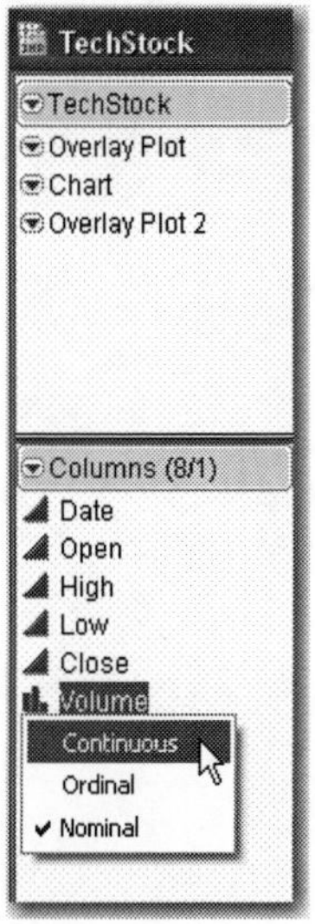

Figure 4.1

Explore

3. Open Graph Builder. Select **Graph > Graph Builder** (see Figure 4.2).

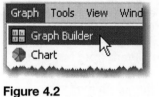

Figure 4.2

4. Notice the columns in your data table appear on the left in the **Select Columns** box (see Figure 4.3).

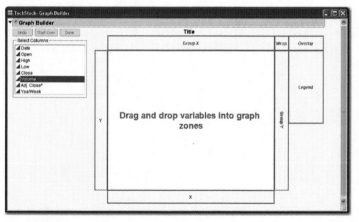

Figure 4.3

5. Now click, hold, and drag the **Adj. Close*** column from the **Select Columns** box to one of the labeled zones located around the graph canvas. You will notice that as soon as the column is dragged into the zone, a graph of that data is immediately expressed. Keeping the mouse button depressed, experiment by dragging the column into different drop zones to see what happens. If you let go of the mouse button, you can always click the **Start Over** button to begin again.

6. Click, hold, and drag the **Adj. Close*** column to the Y drop zone and release the mouse button. You should see a point graph (see Figure 4.4).

7. Now click and drag the **Date** column to the X drop zone. This action immediately produces a scatterplot with dates on the x-axis (see Figure 4.5). Notice that the trend for the close prices is down and a smoothing line is fit to the data.

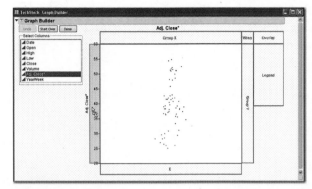

Figure 4.4

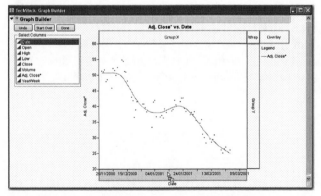

Figure 4.5

Explore

8. Right-click on the points to change the look of the graph (see Figure 4.6). A menu with options to make changes to the graph appears.

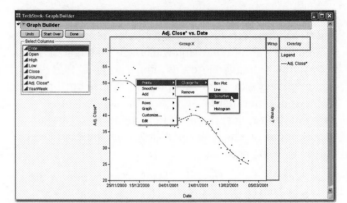

Figure 4.6

9. Now double-click on the **Date** axis. You see a window box of choices (see Figure 4.7). Change the **Tick Label Orientation** to **Vertical**. In **Tick marks and Grid Lines**, add a check for gridline for the **Major** ticks and click **OK**.

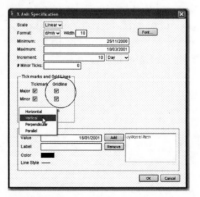

Figure 4.7

10. You should now see a graph with vertical dates and gridlines at the major tick date intervals. Now drag the **Volume** column to the area just above the y-axis label for **Adj. Close*** (see Figure 4.8).

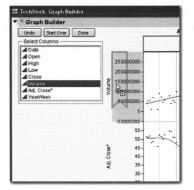

Figure 4.8

11. You will see a graph rendered as in Figure 4.9.

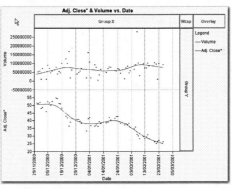

Figure 4.9

Explore

12. Right-click on the **Volume** graph points (see Figure 4.10), and select **Points > Change to > Bar**.

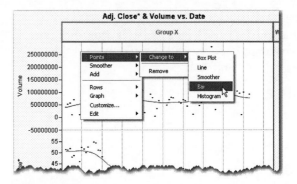

Figure 4.10

13. A graph showing trading volume and adjusted closing prices for the time period is now rendered (see Figure 4.11).

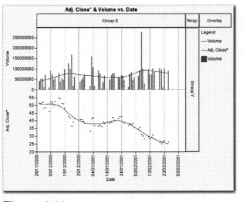

Figure 4.11

14. Now let's include the daily high and low prices in the graph. Drag the **High** column to the location just above the smooth line (see Figure 4.12).

15. A red **High** line is now rendered that represents the maximum price of the stock portfolio by day. Now do the same with the **Low** column by clicking and dragging the **Low** column to the same location. A green **Low** line is now rendered that represents the minimum price for each day.

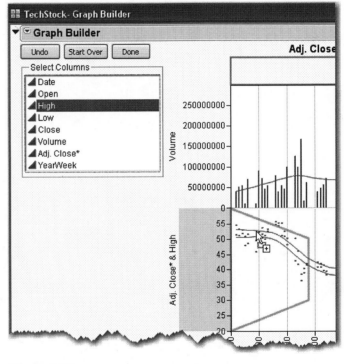

Figure 4.12

Explore

16. To complete the graph, edit the legend. Right-click in the area above the legend and select **Legend Settings** (see Figure 4.13).

Figure 4.13

17. A Legend Settings window appears (see Figure 4.14).

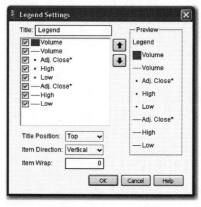

Figure 4.14

18. Click on the blue **Volume** line and remove the check mark for that item. Then click in the **Preview** area of the dialog box. The dialog box should look like Figure 4.15. Click **OK**.

Figure 4.15

19. Now let's add a few last details. We'll change the graph title. Double-click on the graph title and change it to **Tech Stock Sell-Off Fall 2000 to Winter 2001**. Then click **Done**. You should see a graph like that in Figure 4.16.

Congratulations, you have completed your first graph using Graph Builder!

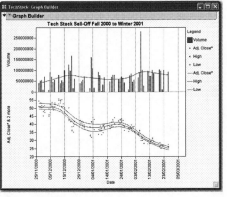

Figure 4.16

Explore

4.2 Using Tabulate

Now that you are familiar with using drop zones to generate a graph, we introduce the Tabulate platform. When you use Tabulate, your goal is to create numerical summaries of your data. This is unlike the Graph Builder platform, where your goal was to create a graph. However, the method of creating a summary of numbers in Tabulate is similar to Graph Builder.

Example 4.2: Movies

We'll use **Movies.jmp** to demonstrate Tabulate. This data table is a listing of popular US movies that were released between 1937 through 2003. The movies are categorized by type, rating, and director and contain earned gross dollar amounts for both the US domestic market and the worldwide market. Note: Dollar amounts are expressed in millions of dollars for both domestic and worldwide markets.

Movie: name of movie

Type: genre/category of movie (for example, comedy, family)

Rating: US movie rating system (for example, general audience [G], adult [R])

Year: year of movie release (for example, 1937)

Domestic $: US domestic revenue in $ earned by the movie in that year

Worldwide $: Worldwide revenue in $ earned by the movie in that year

Director: director of movie

You can find the data set at **Help > Sample Data > Business and Demographics > Movies**.

1. After opening the data table, open **Tabulate** (select **Tables > Tabulate**) (see Figure 4.17).

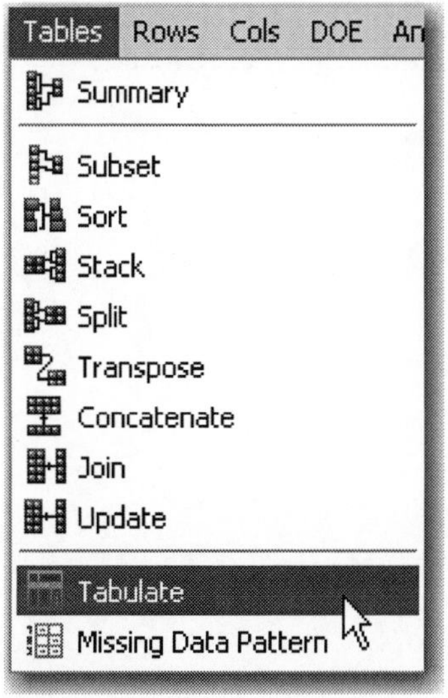

Figure 4.17

Explore

2. The Tabulate platform presents a control panel with your columns, a statistics list, and drop zones for rows and columns (see Figure 4.18).

Figure 4.18

3. Click on both the **Domestic $** and **Worldwide $** columns and drag them to the part of the table labeled **Drop zone for columns**. When you release the mouse button, a dialog box appears. Select **Add Analysis Columns** (see Figure 4.19).

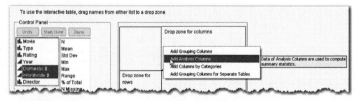

Figure 4.19

4. A sum of movie revenue by domestic $ and worldwide $ appears (see Figure 4.20).

Domestic $	Worldwide $
Sum	Sum
$43,622.40	$84,048.60

Figure 4.20

5. Now drag the **Type** column to the row area (see Figure 4.21).

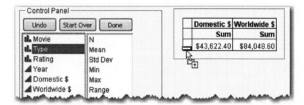

Figure 4.21

6. Release the mouse button and the table now sums the revenue by type for both domestic and worldwide revenue (see Figure 4.22).

Type	Domestic $ Sum	Worldwide $ Sum
Action	$9,224.20	$19,610.10
Comedy	$9,818.00	$15,746.10
Drama	$12,789.80	$24,856.70
Family	$7,575.40	$15,042.80
Mystery-Suspense	$4,031.60	$8,348.60

Figure 4.22

Explore

7. We would like to add a Ratings subgroup to Type of movie. Drag the **Rating** column to the area just to the right of the **Type** column until you see the small plus sign appear, as in Figure 4.23.

	Domestic $	Worldwide $
Type	**Sum**	**Sum**
Action	$9,224.20	$19,610.10
Comedy	$9,818.00	$15,746.10
Drama	$12,789.80	$24,856.70
Family	$7,575.40	$15,042.80
Mystery-Suspense	$4,031.60	$8,348.60

Figure 4.23

8. Release the mouse button and a **Rating** column appears (see Figure 4.24).

		Domestic $	Worldwide $
Type	**Rating**	**Sum**	**Sum**
Action	G	$100.50	$100.50
	PG	$899.00	$1,482.40
	PG-13	$5,732.60	$12,269.00
	R	$2,492.10	$5,758.20
Comedy	PG	$2,825.70	$4,409.70
	PG-13	$4,634.50	$7,498.60
	R	$2,357.80	$3,837.80
Drama	G	$198.70	$390.50
	PG	$5,662.10	$9,579.10
	PG-13	$3,719.90	$8,897.50
	R	$3,209.10	$5,989.60
Family	G	$3,640.10	$7,091.10
	PG	$3,791.10	$7,658.30
	PG-13	$144.20	$293.40
Mystery-Suspense	PG-13	$1,965.20	$4,342.80
	R	$2,066.40	$4,005.80

Figure 4.24

9. Drag the Mean from the **Statistics** column and drop it on one of the **Sum** column headings (see Figure 4.25).

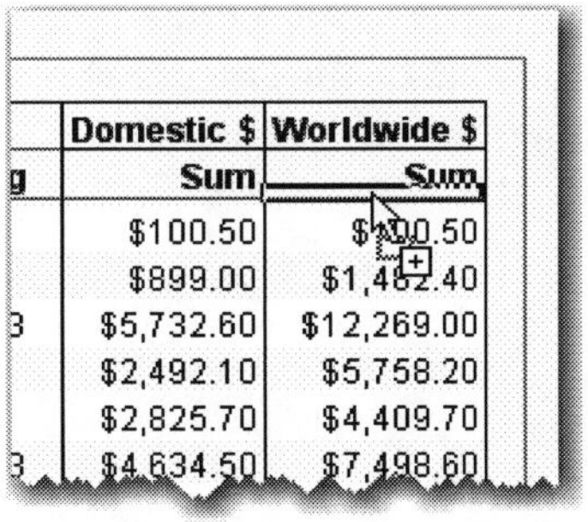

Figure 4.25

10. Then release the mouse button and the **Sum** columns have been transformed to **Mean** columns (see Figure 4.26).

Type	Rating	Domestic $ Mean	Worldwide $ Mean
Action	G	$100.500	$100.500
	PG	$149.833	$247.067
	PG-13	$184.923	$395.774
	R	$138.450	$319.900
Comedy	PG	$141.285	$220.485
	PG-13	$144.828	$234.331
	R	$138.694	$225.753
Drama	G	$198.700	$390.500
	PG	$161.774	$273.689
	PG-13	$232.494	$556.094
	R	$128.364	$239.584
Family	G	$158.265	$308.309
	PG	$180.529	$364.681
	PG-13	$144.200	$293.400
Mystery-Suspense	PG-13	$151.169	$334.062
	R	$129.150	$250.363

Figure 4.26

*Note: You can click the **Undo** button at any time to undo the last action, or you can click **Start Over** to start over again.*

11. Now drag the **Sum** statistic and put it on top of the **Worldwide $** column (see Figure 4.27). Then release the mouse button and **Sum** columns then reappear in the table (see Figure 4.28).

		Domestic $	Worldwide $
Type	Rating	Mean	Mean
Action	G	$100.500	$100.500
	PG	$149.833	$247.067
	PG-13	$184.923	$395.774
	R	$138.450	$319.900
Comedy	PG	$141.285	$220.485
	PG-13	$144.828	$234.331
	R	$138.694	$225.753
Drama	G	$198.700	$390.500
	PG	$161.774	$273.689
	PG-13	$232.494	$556.094

(Palette options: N, Mean, Std Dev, Min, Max, Range, % of Total, N Missing, Sum, Sum Wgt, Variance)

Figure 4.27

		Domestic $		Worldwide $	
Type	Rating	Sum	Mean	Sum	Mean
Action	G	$100.50	$100.500	$100.50	$100.500
	PG	$899.00	$149.833	$1,482.40	$247.067
	PG-13	$5,732.60	$184.923	$12,269.00	$395.774
	R	$2,492.10	$138.450	$5,758.20	$319.900
Comedy	PG	$2,825.70	$141.285	$4,409.70	$220.485
	PG-13	$4,634.50	$144.828	$7,498.60	$234.331
	R	$2,357.80	$138.694	$3,837.80	$225.753
Drama	G	$198.70	$198.700	$390.50	$390.500
	PG	$5,662.10	$161.774	$9,579.10	$273.689
	PG-13	$3,719.90	$232.494	$8,897.50	$556.094
	R	$3,209.10	$128.364	$5,989.60	$239.584
Family	G	$3,640.10	$158.265	$7,091.10	$308.309
	PG	$3,791.10	$180.529	$7,658.30	$364.681
	PG-13	$144.20	$144.200	$293.40	$293.400
Mystery-Suspense	PG-13	$1,965.20	$151.169	$4,342.80	$334.062
	R	$2,066.40	$129.150	$4,005.80	$250.363

Figure 4.28

12. The decimal and display formats can be changed. In the previous example, there are three decimals displayed to the right of the decimal point for the Mean column. We would like to display only two decimal places for values in the table. To adjust the display of decimals, click **Change Format** (see Figure 4.29).

Figure 4.29

13. An additional format panel appears. In the format pane, enter a 2 for the Mean column for both Domestic $ and Worldwide $ so the panel looks as it is shown (see Figure 4.30a). Click the Set Format button. Select the check box for **Use Same Decimal format for all**.

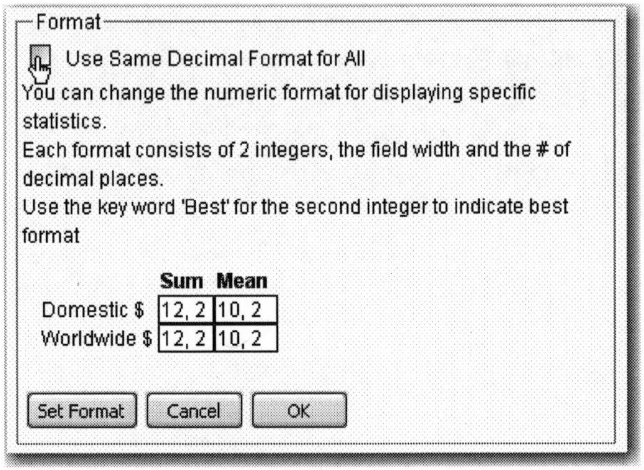

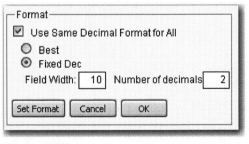

Figure 4.30a

The window changes (see Figure 4.30b). Click **Fixed Dec** and enter a 2 for **Number of Decimals**. Then click **Set Format** and **OK**.

Figure 4.30b

Explore

14. The table is now complete. Click **Done** (see Figure 4.31).

Figure 4.31

15. Recall that additional options are available on the red triangle. Click **Make Into Data Table** (see Figure 4.32) which will produce a JMP data table of the Tabulate results (see Figure 4.33). Tabulate, therefore, is a useful tool for reshaping and reorganizing data for further analysis in addition to producing tabular reports.

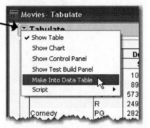

Figure 4.32

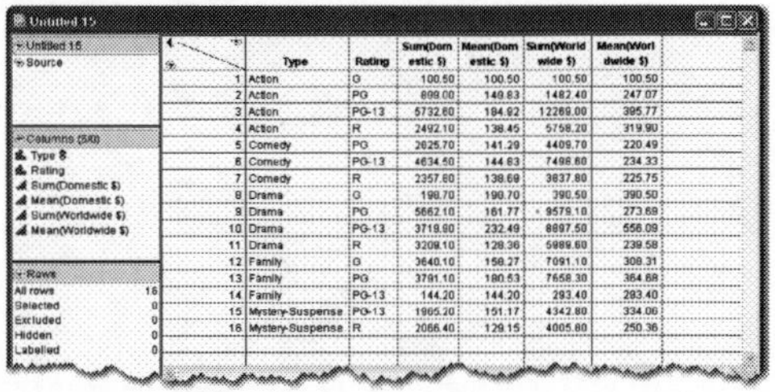

Figure 4.33

Explore

What do you think you can do with the new table in Figure 4.33? Try putting your new skills to work. Use Graph Builder to make a graph out of the data table you just produced (select **Graph > Graph Builder**).

16. Now, press **CTRL** and select the columns **Sum (Domestic $)** and **Sum (Worldwide $)** so that they are both highlighted (see Figure 4.34).

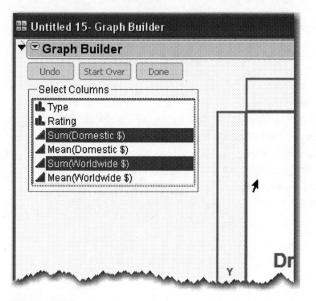

Figure 4.34

17. Drag the two selected columns to the **Y** drop zone (see Figure 4.35).

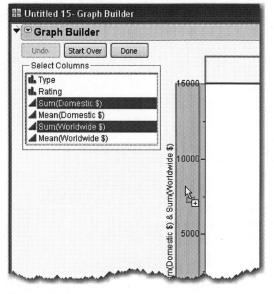

Figure 4.35

Explore

18. A point chart appears with blue and red dots. Right-click in the middle of the points and select **Points > Change to > Bar** (see Figure 4.36).

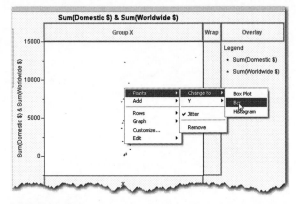

Figure 4.36

19. A bar chart appears showing the worldwide and domestic revenue comparisons (see Figure 4.37).

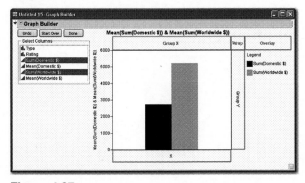
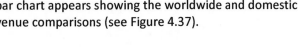

Figure 4.37

20. Now, drag **Type** to the **Group X** drop zone (see Figure 4.38).

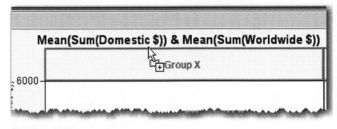

Figure 4.38

21. Now, drag **Rating** to the **Group Y** drop zone (see Figure 4.39).

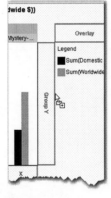

Figure 4.39

Explore

22. The nearly finished graph appears (see Figure 4.40).

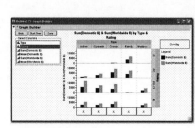

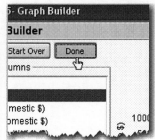

Figure 4.40

Figure 4.41

23. Click **Done** and the finished graph appears (see Figures 4.41 and 4.42).

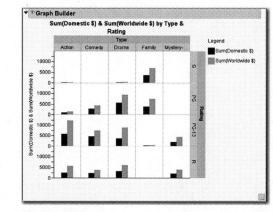

Figure 4.42

4.3 Summary

In this chapter, we took advantage of a key feature of JMP, which is its ability to create graphs and summaries interactively and intuitively in an environment that enables quick exploration and discovery. This chapter introduced you to two tools, Graph Builder and Tabulate.

Graph Builder shows you quick previews of graphs as you drag columns to drop zones, enabling you to flip through choices until you find just the right one to tell the graphical story in your data. Contrast Graph Builder with the methods employed in Chapter 3, where it is assumed that you have a specific graph in mind and need directions to reproduce it.

Tabulate shows you quick previews of table summaries as you drag columns to drop zones. Tabulate enables you to flip through choices until you find the right numerical table summary in your data.

Both Graph Builder and Tabulate employ a method that is like shopping for the right graph or summary and can be combined to tell the best story in your data.

Explore

Chapter 5 • Problem Solving with One and Two Columns

5.1 Introduction

5.2 Analyzing One Column

5.3 Comparing One Column to Another

5.4 Summary

In contrast to earlier chapters that have focused on describing data and producing graphs, this chapter is about getting answers to your questions and making sense of your data. This problem-solving activity is often a process of trial and error, and it does not lend itself to brief descriptive steps. It takes thought and practice to do this well, but JMP is the perfect companion on this journey. This chapter helps you develop some appreciation and basic JMP skills in this problem-solving process.

Just as we discussed in the first chapter, JMP provides a navigation framework that is designed around the workflow of the problem solver. So what do we mean by, "the workflow of the problem solver"? First off, we are talking about the class of problems that are measurable or countable or that already have data. If you need assistance importing or accessing your data, see Chapter 2. By workflow, we are referring to the process by which you analyze data effectively. We have found that when JMP users think about the questions they have of their data and understand how the questions translate to the menu items, they find JMP to be a very intuitive partner in that problem-solving process.

Often multiple questions, or the answer to one question, prompt other questions. JMP is designed to help you answer these follow-up questions quickly. Throughout this chapter, we use simple examples with scenarios that prompt questions one might want to answer with them. We show you how JMP's menus translate to these questions and how the results help you answer them. Just as many real-world problems start with basic questions and understanding and evolve into more complex ones, we start off with the basics here as well.

5.1 Introduction

The transformation of data into information is a process that involves a few basic JMP platforms that we introduce in this chapter, including **Distribution** and **Fit Y by X**. Distribution and Fit X by Y correspond to analyzing the characteristics of one and two columns, respectively, which is the scope of this chapter. The tools are found in the Analyze menu. Table 5.1 outlines the tool name, how many columns it supports, and its statistical terms. Don't worry if you don't recognize the statistical terms and acronyms; we'll present the basic ideas as we go*.

The organization of the items on the **Analyze** menu is the same framework discussed in Section 1.3. Within this framework, we cover just a few menu items but in the process give you access to over 100 statistical methods.

Note: If there is magic to JMP, this is it. First and early questions are answered with the Distribution platform and then further questions are addressed with items further down the menu until you get the answers you need to solve your problem. Thus, you proceed from the simple to just the level of complexity you need to answer your questions as you work your way down (and sometimes back up!) the Analyze menu.

Analyze Menu	How Many Columns	Statistical Terminology
Distribution	Single Column	Univariate methods, histograms, box plots, quantiles, moments, and descriptive statistics and more
Fit Y by X	Two Columns	Bivariate, contingency, logistic, oneway ANOVA, and nonparametrics and more

Table 5.1

---

*Appendix B describes the statistical terms and concepts in this chapter.

5.2 Analyzing One Column

JMP's goal in using this menu framework is to expose you to powerful methods in a logical order that allows you to learn about your data progressively. The order is built into the menu structure.

Figure 5.1 displays the first item on the Analyze menu. It is the Distribution platform. It's first on the menu because this platform answers all of the questions that you should ask early about unknown data.

Another way to remember Distribution as a starting point is that the platform allows you to look at one column at a time and produces results for individual columns. In fact, a branch of statistics is called univariate methods meaning one variable. (In JMP variables of course are referred to as columns as we've mentioned in earlier chapters.) There are dozens of univariate methods, and they are all conveniently arranged in the Distribution platform.

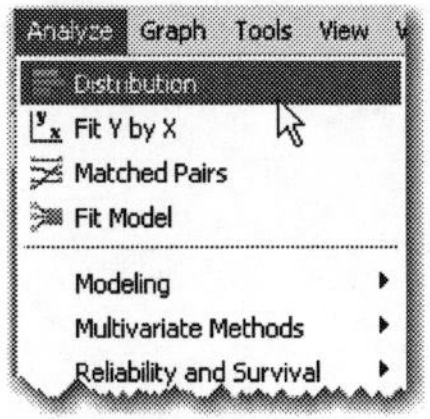

Figure 5.1

Example 5.2: Financial

We will be using the **Financial.jmp** data file to illustrate the steps in this chapter. The data is from the Fortune 500 selected from the April 23, 1990 *Fortune* magazine issue. This data includes columns for:

Type: type of company

Sales($M): yearly sales in millions of dollars

Profit($M): yearly profits in millions of dollars

#emp: number of employees at time of measurement

Profits/emp: profits per employee in thousands of dollars

Assets($Mil.): assets in millions of dollars

Sales/emp: sales per employee in thousands of dollars

Stockholder's Eq($Mil.): stockholder's equity in millions of dollars

You can find the data set at **Help > Sample Data > Business & Demographic > Financial**.

Let's start working our way through an analysis using this process. We'll use this exercise to practice the early questions one asks of unknown data using the Distribution platform in JMP. Let's open a data table on which to base some analysis:

1. Select **Help > Sample Data** (see Figure 5.2).

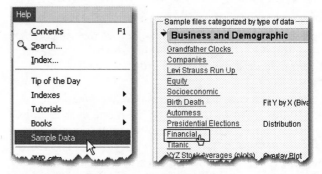

Figure 5.2   **Figure 5.3**

2. Then select **Business and Demographic > Financial** (see Figure 5.3). Some financial performance data is displayed (see Figure 5.4).

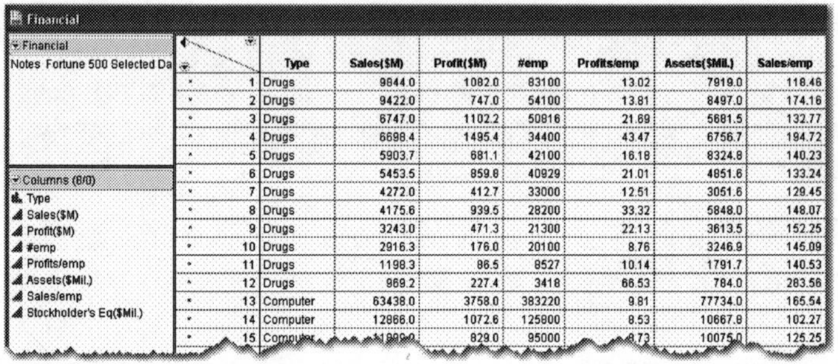

Figure 5.4

Let's assume that your objective is to use this financial data to help you select company stocks to add to your portfolio. Your goal is to select stocks that will maximize the likelihood of positive returns and to use those returns to fund your favorite charity, Rhino Watch. We assume that:

- The most profitable companies will also tend to have the highest positive returns.

- You have sufficient data to make a reasonable prediction.

- Each row represents a company stock. You will need to select about 10 company stocks to sufficiently diversify your portfolio.

- Market conditions for the next 6 months will remain mostly the same for all sectors.

What kinds of questions will you ask to pick company stocks?

Questions That Involve One Column

In this example, you might ask what range of company profits has existed for these stocks. Or, what has been the average (the mean) profit for the companies? How much variability (standard deviation) has there been? Are there some companies whose profits or losses are very extreme (outliers) relative to others? Are these extremes genuine or did someone make a mistake when entering the data (data quality)?

These are the types of questions you might ask of any set of data. These initial questions and many more are all answered with the Distribution platform from the **Analyze** menu.

Note: Time is an important variable to consider with financial data in particular. For the purposes of illustration, however, we have omitted this variable.

Using Distribution to Understand a Column of Data

Let's perform a distribution analysis to answer these early questions for company performance:

1. Select **Analyze > Distribution**.

2. Select **Profits($M)** and **Type** and click **Y,Columns**. A fully populated window now appears (see Figure 5.5).

3. Click **OK**.

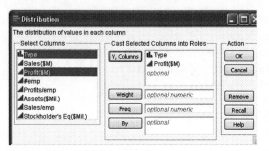

Figure 5.5

You are presented with a result (see Figure 5.6). In this example, you can see several things that might capture your attention. Can you find them? Notice that there are higher, if not extreme, profits among some companies.

You might also notice that some company types in the portfolio are more represented than others. You can see that Oil has the most companies and Beverages the least because of the size of the bars in the Type distribution graph.

Note: Did you know that graphs and tables of numerical results appear together in report windows by design? Recent research indicates that 70% of learning is done visually. Brain studies indicate that people tend to favor either pictures or numbers, so both output types are almost always displayed in report windows in JMP to help both kinds of learning.

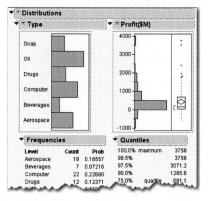

Figure 5.6

Now let's try something new to help us answer some important questions:

4. Click and draw a box with your pointer on some of the highest profit entries (see Figure 5.7). Just take the pointer, left-click and drag down over the dots in the profit graph, and a rectangle appears.

5. Notice that as you draw the box, the square dots become highlighted and appear as larger points.

6. Notice also the graph bars on the left for company type, where certain types of company bars have started to turn dark green*, including Oil, Computer, and Beverages.

7. This means that companies with the greatest profits are also those mostly associated with the oil sector. Notice the tiny slivers of Computer and Beverages are turning dark green, which indicates a small number of highly profitable companies are in these two business sectors.

*Color references used in this chapter refer to what you will see on your screen using the standard default settings in JMP.

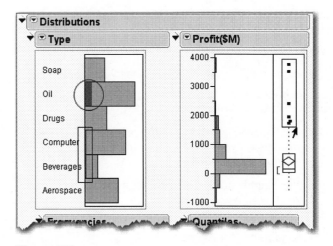

Figure 5.7

*Key Concept: **Dynamic Linking of Graphs and Data.** Nearly all graphs that appear in report windows in JMP are tied directly to any other displayed graphs AND to rows in the corresponding data table. By selecting graphical attributes, you see those corresponding values represented in other graphs AND highlighted in the data table.*

8. Now select the **Window** menu from the top and select **Financial**.

This brings the data window to the foreground.

9. Now scroll down to around row 35 (see Figure 5.8).

Do you see how the highlighted rows match those same points you highlighted by drawing the box?

Note: The highlighting is what we refer to as JMP's dynamic linking, which allows you to link graphs to data, data to graphs, and graphs to graphs.

Because these highlighted rows show high profits, let's mark them so we can always find them.

10. Select **Rows > Colors**, and then select the color red from the color palette (see Figure 5.9).

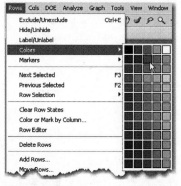

	Type	Sales($M)	Profit($M)	#emp	Profits/
33	Computer	784.7	89.0	4708	
34	Computer	709.3	41.4	5000	
35	Oil	86656.0	3510.0	104000	
36	Oil	50976.0	1809.0	67900	
37	Oil	32416.0	2413.0	37067	
38	Oil	29443.0	251.0	54826	
39	Oil	24214.0	1610.0	53648	
40	Oil	21703.0	1405.0	31338	
41	Oil	17755.0	965.0	53610	
42	Oil	15905.0	1953.0	26600	
43	Oil	12492.0	219.0	21800	
44	Oil	10417.0	260.0	17286	
45	Oil	9927.0	98.0	21600	
46	Oil	8685.6	170.1	13232	
47	Oil	8016.6	86.2	37800	
48	Oil	5589.0	476.3	8740	
49	Oil	4941.2	161.0	3300	
50	Oil	3122.0	156.0	7942	

Figure 5.8

Figure 5.9

11. Select **Rows > Markers**, and select the marker type **X** (see Figure 5.10).

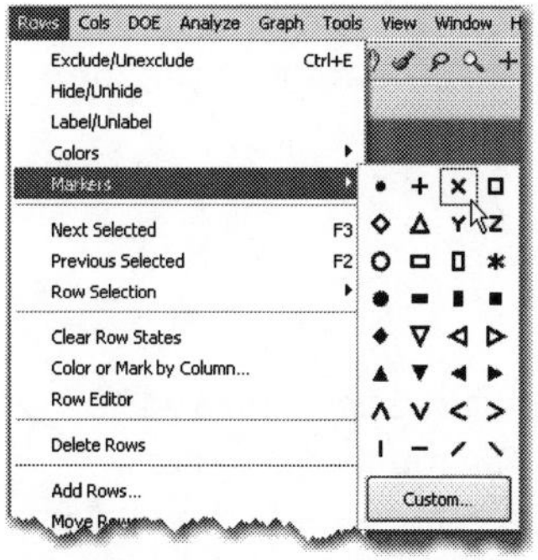

Figure 5.10

These rows in the data table are now marked red and appear with a marker type of **X**, as shown in the row number column of the data table (see Figure 5.11).

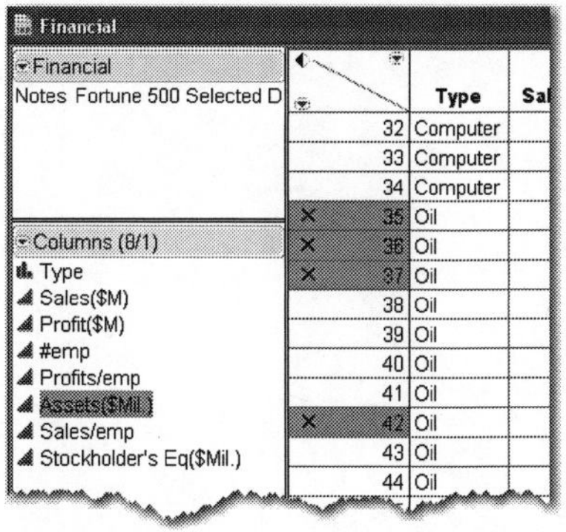

Figure 5.11

12. Now select **Window** and bring the **Distributions** result to the foreground. Focus on the Distributions result for **Profit($M)** (see Figure 5.12).

The distribution graph for profit also shows the red X markers as well as a box plot. The top X represents a company with over $ 3.7 billion in profits and the bottom of the range is a company with a loss of around $1 billion (see Figure 5.12).

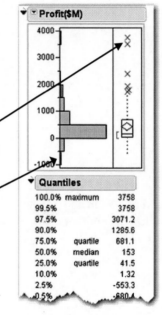

Figure 5.12

If a box plot or any graphical or numerical result is unfamiliar to you, use the question mark tool from the **Tools** menu to find out what the item in question is and can say about your results.

13. Select the question mark **(?)** from the toolbar. The pointer changes to a question mark (see Figure 5.13).

14. Then move the question mark on top of the item you're unfamiliar with and click on that item (see Figure 5.14). In this example, it is the item next to the distribution graph.

Figure 5.13

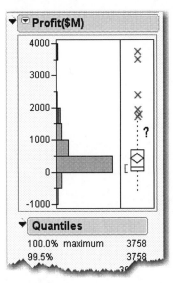

Figure 5.14

Problem Solving I

The section of the documentation associated with the outlier box plot appears automatically (see Figure 5.15). Don't forget to scroll down, because sometimes the topic of interest is a little below where you landed in the documentation.

*Key Concept: Unfamiliar graphics, geometry, and results are explained using the context-sensitive question mark tool. Just select the question mark tool (?) from the **Tools** menu and click on the part of the result you want to learn about. You can learn statistics at the same time you explore your data!*

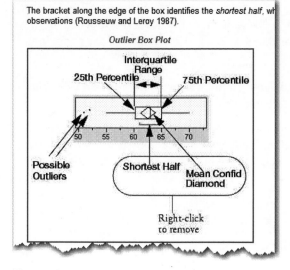

Figure 5.15

Summary of 5.2:   Answers the Distribution Platform Provides

By observing the distribution, you found that the range of profits for the companies was mostly on the high side, topping out with a highly profitable company at over $3.7 billion. On the bottom of the range was a very unprofitable company at approximately -$1 billion (Figure 5.14).

You may have noticed that the average, or mean, profits for all companies was $426 million.

The distribution results indicated that the variability, or standard deviation, of profits was about $717 million. Most of the profitable companies fall within this range around the mean.

You found that most of the extreme values were on the high side. By identifying them, you determined that most of these companies came from the oil sector and a couple of other industry types.

5.3 Comparing One Column to Another

Building upon the Distribution platform, the answers to questions you asked about one column have motivated you to ask further questions about the relationship between two columns. For example, what might be the relationship of profits to other columns in the data table? In fact you have already identified an interesting relationship between the company type and its profits using dynamic linking as some types of companies are associated with higher profits. This visual relationship implies something is happening with profits that includes more than just one column. You identified a relationship between the Profits column by another column called Type. This implies that there may be a relationship between at least two columns that we'll explore in this section.

The second item under the **Analyze** menu is **Fit Y by X**. It is designed to explore relationships between one column and another column. These are sometimes called *bivariate relationships* (meaning *two variables*). You might have noticed a pattern developing for items under the **Analyze** menu. The first item is **Distribution** and is useful for looking at one column at a time. The second item is **Fit Y by X** for looking at two columns in relationship. Figure 5.16 provides a description of each menu item.

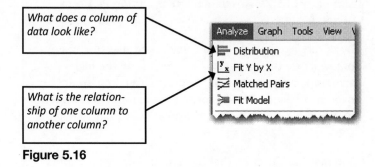

Figure 5.16

Let's continue with our example and perform a Fit Y by X analysis. We already suspect that there are interesting profit differences among the different company types, and we've discovered a few by just looking at one column at a time using the Distribution platform and dynamic linking. We'll now formalize that inference.

The relationship we want to explore includes the **Profit($M)** column and the **Type** column:

1. Select **Analyze > Fit Y by X** (see Figure 5.17).

2. In the window (see Figure 5.18), select **Profit($M)** and click **Y, Response.**

3. Select **Type** and click **X, Factor.**

4. Click **OK.**

Before we continue with the example, let's examine that last Fit Y by X window. We'll focus on the preview or circled area (see Figure 5.18).

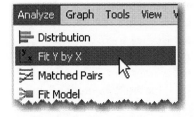

Figure 5.17

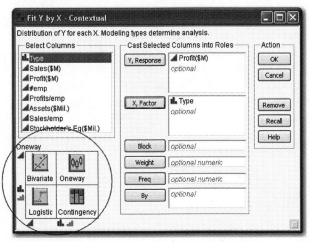

Figure 5.18

Notice that the modeling type of the column (continuous, nominal, ordinal) that you cast into a role determines the type of analysis that is produced (see Figure 5.19). Each modeling type has its corresponding icon. These modeling types and their icons are described in Section 2.3. In this case, we have selected **Profit($M)** for the **Y, Response** (the vertical axis), which is continuous, and **Type** for the X, Factor (the horizontal axis), which is nominal. The preview works like a 2-by-2 table and tells you that some kind of graph will be produced. For now, don't worry about terms like *oneway* in the preview. If you want to learn about the result, you can put the question mark tool on the result after generating it. Just note that the picture previews are there so you can get an idea of what kinds of analyses will be produced when you cast certain types of columns into roles in the **Fit Y by X** platform.

When you clicked **OK**, a oneway analysis appeared (see Figure 5.20).

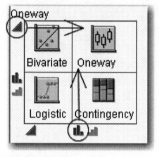

Figure 5.19

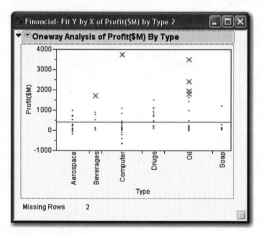

Figure 5.20

The Fit Y by X (two-column) analysis confirms what we started to observe earlier. We see that the highest profits have come from a mix of mostly oil companies, one computer company, and one beverage company. These are conveniently marked by Xs from a previous step.

Based on the observed best performers among the company types, we might start asking more complex questions like these:

> **Questions Answered with Two-Column Analysis**
>
> -What are the differences in profits among the company types?
>
> -How big and in what direction (negative or positive) are these differences?
>
> -Should I choose more companies from one particular company type than from another type?

These questions and others can be answered by Fit Y by X because they involve two columns. How? As you make selections from the choices in the **Analyze** menu, the types of questions you start asking at each step are anticipated for you.

You might notice the hot spot (or small red triangle) associated with the results you have generated (see Figure 5.21).

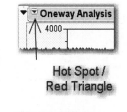

Figure 5.21

As we introduced earlier, *hot spots* anticipate questions you might have at any stage of analysis and have been carefully placed on the menu in the context of your analysis.

Let's answer some of the questions you have using the choices on the menu for the oneway analysis:

-What are the differences in profits among the company types?

You can already see some of these differences in the graph. Let's further quantify the differences in profits by company type.

From the hot spot, select **Means and Std Dev** (see Figure 5.22).

An additional table appears below your graph (see Figure 5.23).

Using the table, we can now see that the highest average profits (mean) were obtained from the 12 companies represented in the drugs category. The second highest average profits were obtained from the oil category, and there were 26 companies represented.

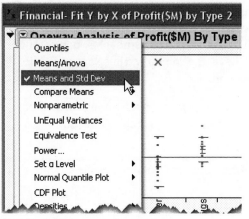

Figure 5.22

Larger samples can help
spread risk.

Means and Std Deviations

Level	Number	Mean	Std Dev	Std Err Mean	Lower 95%	Upper 95%
Aerospace	18	238.678	310.497	73.18	84.3	393.1
Beverages	7	590.886	608.206	229.88	28.4	1153.4
Computer	22	257.259	879.254	187.46	-132.6	647.1
Drugs	12	690.075	429.988	124.13	416.9	963.3
Oil	26	616.938	824.741	181.36	243.4	990.4
Soap	10	205.690	360.619	114.04	-52.3	463.7

Figure 5.23

Let's do a check on what we have learned so far. In Distribution, we learned that companies with the highest profits are likely to be in the oil category. From Fit Y by X, we learned that the highest average profits are found overall in the drug category and the next highest average profits are found in the oil category. Remember, you are trying to accumulate just enough knowledge to make your decisions. If you are keeping score, Oil is showing up as a good performer by at least two measures. The most extreme profits are coming from some of these oil companies and overall averages are also high for Oil.

Did you also notice the number of oil companies represented? This is another vote for Oil to be represented in our selection of company stocks because many highly profitable stocks are from this category. The risk associated with one oil company going down is hedged by having many profitable ones in the oil category.* If you are still keeping score, Oil looks better by three measures now.

We don't really have numerical measurements of range yet to help answer the question: "How big and in what direction (negative and positive) do these differences go?"

Where might we go to answer this question? Yes, it's in the red triangle for the oneway result.

1. Select the **red triangle** for **Oneway**.

2. Select **Quantiles** (see Figure 5.24).

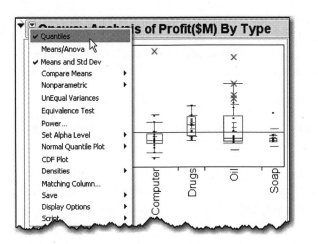

Figure 5.24

> **Quantiles** are values that divide an ordered set of continuous data (from smallest to largest) into equal proportions. These proportions correspond to percentile break points. See "Quantiles" in Appendix B for more information.

This is true when the companies in the category are not highly correlated on the fundamentals. Seek professional advice when making investment decisions.

A Quantiles table appears (see Figure 5.25). This table appears in the context of your initial analysis, and box-and-whisker plots are added to your graph illustrating what the quantiles are indicating.

The box encompasses the 25th through 75th percentiles.

The whisker lines stretch out to the 10th and 90th percentiles on the corresponding sides of the box.

▼ Quantiles							
Level	Minimum	10%	25%	Median	75%	90%	Maximum
Aerospace	-187	-91.15	27.3	162.1	362.175	758.71	973
Beverages	13.1	13.1	71.7	514.5	901.4	1723.8	1723.8
Computer	-680.4	-574.8	-29.7	70.15	352.975	999.52	3758
Drugs	86.5	113.35	273.725	714.05	1046.375	1377.44	1495.4
Oil	-30.5	18.24	45.325	158.5	1075	2091	3510
Soap	14.4	15.58	28.6	89.35	174.25	1113.4	1206

Figure 5.25

What type of company has the worst profit by examining the **Minimum** column? What type of company has the best profit in the **Maximum** column? Which has the lowest value for the **Maximum** column?

You can make it easier to see. Try this:

1. Right-click on the **Quantiles** table (see Figure 5.26).

2. From the submenu, select **Sort by Column** (see Figure 5.26).

▼ Quantiles							
Level	Minimum	10%	25%	Median	75%	90%	Maximum
Aerospace	-187	-91.15	27.3	162.1	362.175	758.71	973
Beverages				14.5	901.4	1723.8	1723.8
Computer	-6			0.15	352.975	999.52	3758
Drugs				4.05	1046.375	1377.44	1495.4
Oil				58.5	1075	2091	3510
Soap				9.35	174.25	1113.4	1206

Submenu:
- Table Style ▸
- Columns ▸
- Sort by Column...
- Make Into Data Table
- Make Combined Data Table
- Make Into Matrix

Figure 5.26

3. Select **Minimum** and click **OK** (see Figure 5.27).

4. The Profit table now appears sorted by the **Minimum** column (see Figure 5.28).

You can now see that the worst loss (or minimum of profit) is in the Computer category at -680.4. Can you find the biggest profits in the **Maximum** column? Yes, it's also in the Computer category at 3758. They are circled for you.

Now you can start asking yourself about your risk tolerance. Review the oneway graph you produced, the mean and standard dev, and the Quantiles table. If you are conservative, what types of companies would you pick to assure profits? Wouldn't you choose only companies that show no negative profits?

Which ones are those? They are Drugs, Soap, and Beverages.

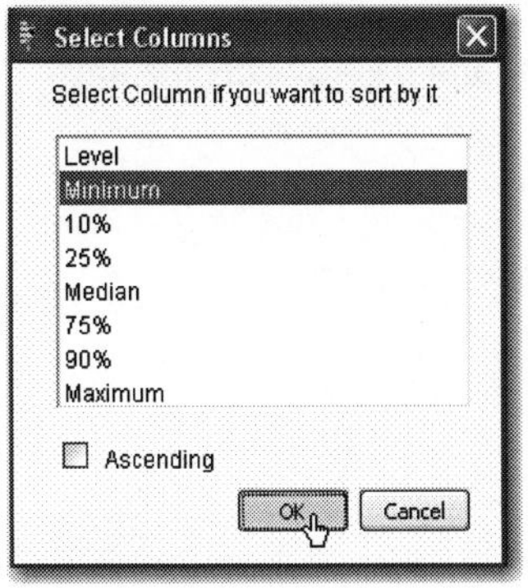

Figure 5.27

Level	Minimum	10%	25%	Median	75%	90%	Maximum
Drugs	86.5	113.35	273.725	714.05	1046.375	1377.44	1495.4
Soap	14.4	15.58	28.6	89.35	174.25	1113.4	1206
Beverages	13.1	13.1	71.7	514.5	901.4	1723.8	1723.8
Oil	-30.5	18.24	45.325	158.5	1075	2091	3510
Aerospace	-187	-91.15	27.3	162.1	362.175	758.71	973
Computer	-680.4	-574.8	-29.7	70.15	352.975	999.52	3758

Figure 5.28

Higher profits appear in other categories, but more of those companies show higher variability and negative profits and, therefore, present more risk.

5. Finally, you need to save the data table **Finance.jmp.** You will need this data table for the next chapter.

Summary of 5.3:   What You Learned Comparing One Column to Another

You may have noticed that the highest mean (average) profits are found overall in the 12 companies in the Drug category and the next highest mean profits are found in the 26 companies in the oil category.

You found that the most negative profits were in the computer category as well as the most profitable ones.

You found that picking companies depends on risk tolerance. If you are conservative, you might pick Drugs, Soap, and Beverages. Higher profits might be achieved elsewhere, but they are in company types where there is higher variability or risk and some of these companies have negative profits.

5.4 Summary

This chapter has presented an approach to problem solving that is unique to JMP. The approach underscores the progressive nature of problem solving that tends to build from simple descriptions of one column of data to relationships between two columns of data, leading to an understanding of the data.

The process of learning about data using JMP tends to start slowly, increases rapidly, and then reaches understanding. Also, the process does not go simply in one direction as you move from one column to two columns. As discoveries are made between two columns, confirmation and further analysis can be made with simple one-column tools like Distribution from which our journey began.

Marking rows with colors and markers helps certain groups stand out in subsequent analyses. This visual identification further speeds discovery and effective communication of results.

Because many real-world problems involve multiple variables or columns, you will learn in the next chapter that this increased complexity is easily handled by JMP. We will build upon our example and the basic analyses introduced in this chapter to explore multivariable relationships. In the next chapter, we introduce several tools including the Partition platform, the Data Filter, and the Prediction Profiler to help you explore and discover deeper insights in your data.

Chapter 6 • Problem Solving with Multiple Columns

6.1 Introduction
6.2 Comparing Multiple Columns
6.3 Filtering Data for Insight
6.4 Model Fitting, Visualization, and What If Analysis
6.5 Summary

The previous chapter focused on problem solving when you have one or two columns of interest. This chapter builds on the previous chapter's framework for problem solving and introduces a method to understand multiple column and group relationships.

We introduce advanced analytical methods including Partition, Fit Model, and Prediction Profiler to help us understand the multi-column/variable relationships in the data. Most real-world problems involve more than one- or two-column relationships so the Partition approach is warranted.

We also introduce the Data Filter for real-time slicing of the data. The Data Filter enables easy inference testing and hypothesis development by excluding, hiding, and marking selected observations across any number of column ranges or groups.

The Fit Model platform includes several modeling personalities, including Multiple Regression. Unlike the two-column analysis (one Y and one X) we introduced in the last chapter, Multiple Regression allows you to employ additional X columns. We briefly introduce this topic at the end of the chapter, along with a very powerful model visualization tool, the Prediction Profiler.

Problem Solving II

6.1 Introduction

As we explained in the previous chapter, the transformation of data into information is a process that involves a few basic JMP platforms including Distribution and Fit Y by X, which are found in the **Analyze** menu. Table 6.1 provides a review of the tool name, how many columns it supports, and its common statistical identity. The process of discovery starts with univariate/one-column analysis using the Distribution platform to explore the single-column/variable properties and proceeds to bivariate/two-column relationships using the **Fit Y by X** platform. As in life itself, these analyses can often lead to more complex questions involving more than two columns, which is the subject of this chapter.

In Table 6.1, we add Fit Model and Modeling to the items on the Analyze menu column. Partition and Multiple Regression are associated with these menus and are added to the table framework. These methods support multivariable or multi-column relationships.

Analyze Menu	How Many Columns	Statistical Terminology
Distribution	Single column	Univariate methods, histograms, box plots, quantiles, moments, and descriptive statistics, and more
Fit y by x	Two columns	Bivariate, contingency, logistic, oneway ANOVA, nonparametrics, and linear regression, and more
Fit Model Modeling	Multi-Column	Partition, Data Mining, Multiple Regression

Table 6.1

Remember, JMP's goal in using this menu framework is to introduce you to methods in a logical order, allowing you to learn about your data progressively, without forcing you to first understand statistical jargon or the conditions required to implement them. The process of digging deeper into your data is built into the menu structure.

6.2 Comparing Multiple Columns

What remains to be answered is whether there are important multiple-column relationships (looking at more than two columns) that might be present in the data. Complex products or systems often have these relationships. While these multiple-column relationships are not always present in the questions you are trying to answer, knowing that you have access to methods to identify them should help. In fact, sometimes to see the forest for the trees, you need these methods so that you can focus on only those columns that drive your learning from the data.

Example 6.2: Financial

We will be using the **Financial.jmp** data file to illustrate the steps in this chapter. The data is from companies in the Fortune 500, selected from the April 23, 1990 *Fortune* magazine issue. This data includes columns for:

Type: type of company

Sales($M): yearly sales in millions of dollars

Profit($M): yearly profits in millions of dollars

#emp: number of employees at time of measurement

Profits/emp: profits per employee in thousands of dollars

Assets($Mil.): assets in millions of dollars

Sales/emp: sales per employee in thousands of dollars

Stockholder's Eq($Mil.): stockholder's equity in millions of dollars

You can access this data at **Help > Sample Data > Business & Demographic > Financial.**

You should have the **Financial.jmp** data table open from the previous chapter. If you don't, then open it: **Help > Sample Data > Business & Demographic > Financial** (see Figure 6.1).

Preparing Data Using Recode

Before we perform an analysis, let's create a new column that simply identifies each company as being profitable or not (we will in effect be making profit a new nominal column*). Our intention is to identify profitable companies based on all the other columns. We want to pick companies whose profits will be sufficient to provide us with solid returns, so we want to identify companies whose profits are above $10 million and store them in a new column. The $10 million mark is an arbitrary break point (it could be any value), but it is based on the idea that companies with higher profits in general will probably be larger and less volatile. For purposes of illustration, we will use $10 million as a reasonable and conservative strategy at this point.

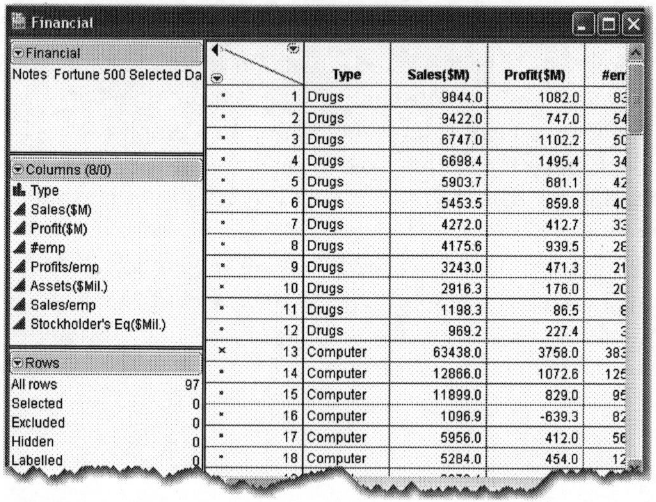

Figure 6.1

Note: We do not need to change the profit column to a nominal type to use Partition. Output columns like profit can be either continuous or nominal in the Partition platform as well as any input columns. We made this change only to simplify the interpretation of our analysis for illustration purposes only.

Let's make **Profits($M)** a nominal column:

1. Select **Rows**.

2. Select **Row Selection**.

3. Select **Where**… (see Figure 6.2). This generates the Select rows window (see Figure 6.3).

4. Select **Profits($M)**.

5. Select **is greater than**.

6. Enter **10**. (The data is expressed in millions.)

7. Click **Add Condition**.

8. Click **OK**.

Figure 6.2

Figure 6.3

Rows that meet the condition of Profits being greater than $10 million are now highlighted in the data (see Figure 6.4).

	Type	Sales($M)	Profit($M)	#emp	Profits/emp	Assets($MIL)
11	Drugs	1198.3	86.5	8527	10.14	1791.7
12	Drugs	969.2	227.4	3418	66.53	784.0
13	Computer	63438.0	3758.0	383220	9.81	77734.0
14	Computer	12866.0	1072.6	125800	8.53	10667.8
15	Computer	11899.0	829.0	95000	8.73	10075.0
16	Computer	1096.9	-639.3	82300	-7.77	10751.0
17	Computer	5956.0	412.0	56000	7.36	4500.0
18	Computer	5284.0	454.0	12068	37.62	2743.9
19	Computer	3078.4	-424.3	28334	-14.97	2725.7
20	Computer	2959.3	252.8	31404	8.05	5611.1
21	Computer	2952.1	-680.4	18000	-37.80	1860.7
22	Computer	2876.1	333.3	9500	35.08	2090.4
23	Computer	2153.7	153.0	8200	18.66	2233.7
24	Computer	1769.2	60.8	10200	5.96	1269.1
25	Computer	1643.9	118.3	9548	12.39	1618.8
26	Computer	1517.9	-230.2	•	•	•
27	Computer	1382.3	0.3	2900	0.10	1076.8
28	Computer	1324.3	-119.7	13740	-8.71	1040.2

Figure 6.4

9. Now select **Rows > Row Selection > Name Selection in Column...** (see Figure 6.5).

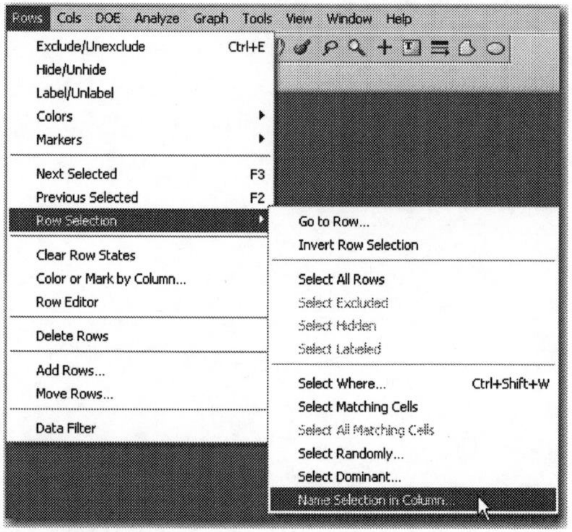

Figure 6.5

10. Type the column name and labels for the **Selected** and the **Unselected** rows exactly as shown (see Figure 6.6) and click **OK**.

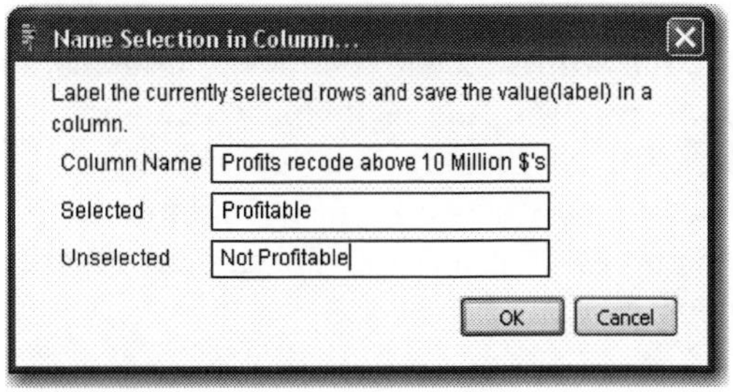

Name Selection in Column...

Label the currently selected rows and save the value(label) in a column.

Column Name	Profits recode above 10 Million $'s
Selected	Profitable
Unselected	Not Profitable

OK   Cancel

Figure 6.6

A new column is created that identifies profitable and not profitable (by our definition) companies for each row in the table, with a dividing point of $10 million and higher (see Figure 6.7). We now have all the information we need to explore the relationship of profitable and not profitable companies vs. all the other columns because we've made a new column that identifies each company by our definition as either profitable or not profitable*. This is called *recode*. Not having to write any code to do this is very handy. (JMP recode is further explained in Section 2.4.) Now we can move on with the analysis.

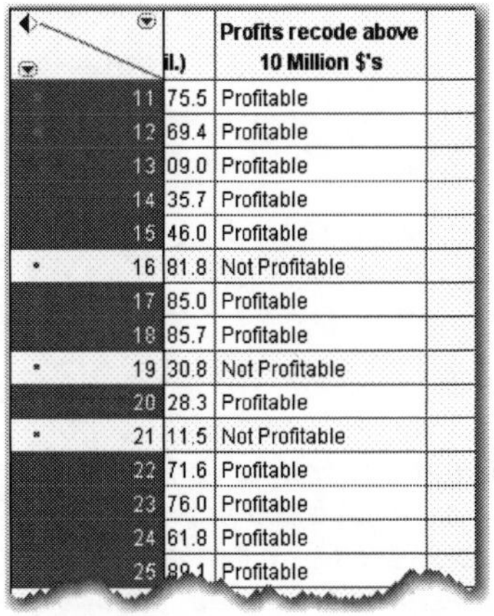

		Profits recode above 10 Million $'s
	il.)	
11	75.5	Profitable
12	69.4	Profitable
13	09.0	Profitable
14	35.7	Profitable
15	46.0	Profitable
16	81.8	Not Profitable
17	85.0	Profitable
18	85.7	Profitable
19	30.8	Not Profitable
20	28.3	Profitable
21	11.5	Not Profitable
22	71.6	Profitable
23	76.0	Profitable
24	61.8	Profitable
25	89.1	Profitable

Figure 6.7

*Note that we will hereafter refer to this new column as "Profit recode."

Problem Solving with Multiple Columns • Comparing Multiple Columns

Consider the data table in Figure 6.8. More than one column might be related to the **Profits recode** column: for example, **Sales($M)**, **#emp**, **Assets($M)**. This might imply that more than two columns are required for the analysis. Therefore, we cannot use only Fit Y by X. There are many methods to investigate this kind of relationship. Let's look at one of these with a tool called Partition.

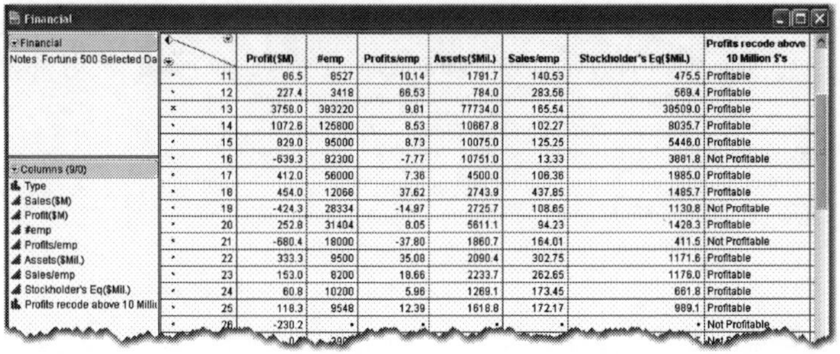

	Profit($M)	#emp	Profits/emp	Assets($Mil.)	Sales/emp	Stockholder's Eq($Mil.)	Profits recode above 10 Million $'s
11	86.5	8527	10.14	1791.7	140.53	475.5	Profitable
12	227.4	3418	66.53	784.0	283.56	569.4	Profitable
13	3758.0	383220	9.81	77734.0	165.54	38509.0	Profitable
14	1072.6	125800	8.53	10667.8	102.27	8035.7	Profitable
15	829.0	95000	8.73	10075.0	125.25	5446.0	Profitable
16	-639.3	82300	-7.77	10751.0	13.33	3881.8	Not Profitable
17	412.0	56000	7.36	4500.0	106.36	1985.0	Profitable
18	454.0	12068	37.62	2743.9	437.85	1485.7	Profitable
19	-424.3	28334	-14.97	2725.7	108.65	1130.8	Not Profitable
20	252.8	31404	8.05	5611.1	94.23	1428.3	Profitable
21	-680.4	18000	-37.80	1860.7	164.01	411.5	Not Profitable
22	333.3	9500	35.08	2090.4	302.75	1171.6	Profitable
23	153.0	8200	18.66	2233.7	262.65	1176.0	Profitable
24	60.8	10200	5.96	1269.1	173.45	661.8	Profitable
25	118.3	9548	12.39	1618.8	172.17	989.1	Profitable
26	-230.2	•	•	•	•	•	Not Profitable

Figure 6.8

Note: You might find these multi-way relationships by simply (but laboriously) producing distributions or examining output from Fit Y by X. We use Partition because it is a powerful and quick method for finding these relationships when you have more than two potential relationships in your data. Partition results are usually easier to interpret as well.

Mining Data Using Partition

The Partition platform allows you to determine which columns in a data table most influence or predict the outcome of another column (profits, in our example). It achieves this by searching all of your selected X columns and finding a set of splits or subgroups within them whose values best predict your Y value. These splits (or partitions) of the data are done recursively (starting with the best predictor), forming a tree of decision rules until the desired fit is reached.

A decision tree is essentially a ranked set of predictors (such as type or assets, in our example) placed in a hierarchical tree structure according to the strength of their relationship to the column of interest (profits or Y, in our example). Items higher in the tree have more influence than items lower in the tree.

In our example, Profits recode is our column of interest and we want to determine which of the remaining columns in our data table most influence it. The Partition method introduced here is sometimes termed data mining because you are mining relationships in your data to find which columns most relate to or predict profits.

To put this in the language of our example, we want to understand the relationship of **Profits recode** (the output or Y column) to all the other columns (input X columns). See Figure 6.9.

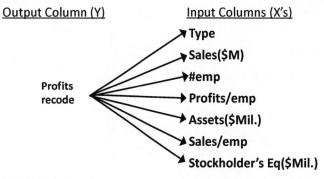

Figure 6.9

Note: The Partition technique is introduced here to enable quick exploration of your data. It is an advanced analytical technique. It is recommended that you familiarize yourself with the underlying analysis concepts by using the question mark tool to call up the relevant chapters in the documentation. Because these tools often require greater sophistication in their interpretation, we recommend that you seek out experienced data analysts for assistance when necessary.

To explore relationships of more than two columns, we will move further down the **Analyze** menu to the **Modeling** menu (see Figure 6.10).

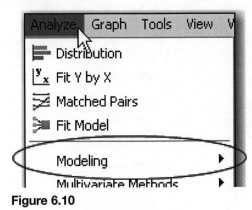

Figure 6.10

Because we have already explored profits compared with the type of company, we might now ask other questions.

These are just a few of the questions that can be answered using the Modeling platforms.

Framing Our Analysis

Questions That Involve Multiple Columns

-What are the best predictors of profit?

-Are there relationships between profits and any of the other columns?

-Where are the biggest differences found between company profits?

-Where are there no differences?

-Is there some way to use multiple columns to pick the most profitable companies?

Problem Solving II

Let's run the Partition platform:

1. Select **Analyze > Modeling > Partition** (see Figure 6.11).

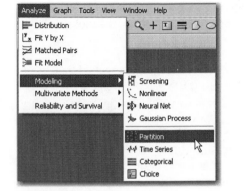

Figure 6.11

2. Select **Profits recode** for **Y, Response** (see Figure 6.12).

3. And for **X, Factor**, select **Type**, **Sales**, **#emp**, **Assets**, **Sales/emp**, and **Stockholder's Eq** (see Figure 6.12).

4. Click **OK**.

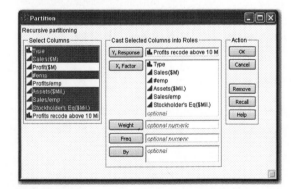

Figure 6.12

Problem Solving II

You now see profits for all the companies represented by a random pattern of dots from left to right in the mosaic plot (see Figure 6.13).

The solid horizontal black line through the range of 0.00 and 1.00 on the vertical axis is the dividing point we set for companies that have at least $10 million in profit.

The companies that are above the line are the profitable ones and those below are the unprofitable ones. So, looking at the line on the mosaic plot, about 13% of the companies are not profitable and about 87% are profitable.

You can see those very high values (outliers) to which we assigned a red X marker from the previous chapter. We saved these in the table as row properties previously. If you don't see the red X's, it's ok. We can go forward without them.

Now let's try something.

5. Click **Color Points**. It's the button next to **Split** and **Prune** (see Figure 6.13).

This identifies all profitable companies with blue markers and all unprofitable companies with red markers. Remember these colors. They are a visual guide to interpreting the results.

Notice two buttons are circled: **Split** and **Prune** (see Figure 6.13).

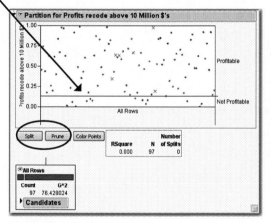

Figure 6.13

Remember that the point of this analysis is to understand the multiple-column relationships between profits and the other columns in the data table.

6. Select **Split** and new output appears (see Figure 6.14).

What happened? Partition went through all the columns you selected as X's and found that the best predictor of difference* in profits was type of company. Remember, you also included a half-dozen other columns in your model, but the first to be selected was the Type column. Also note that your high performers from the previous chapter's work show up as red X's (that is, if you saved them in the previous chapter). These should show up as profitable as you discovered them using the Distribution platform. It's a good confirmation that Partition found them also!

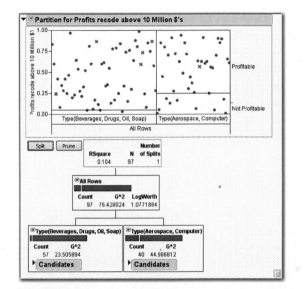

Figure 6.14

*This difference is important and is expressed by finding the column whose categories (or difference in means, in the case of a continuous column) are most different as they relate to our Y, thus allowing us to predict our Y from our split X values. Want to know more about the analytical method used? Go to **Help > JMP Stat and Graph Guide > Chapter 37 > Partition**.

Let's express the splits as probabilities, so go to the hot spot/ red triangle: ——

7. Select **Display Options**.

8. Select **Show Split Prob** (see Figure 6.15).

9. Select **Split** again. Another branch apears on the report (Figure 6.16).

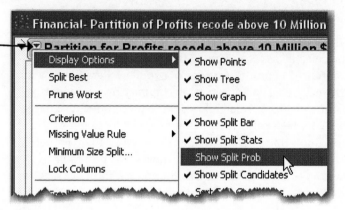

Figure 6.15

*A Few Notes on Partition Trees: Partition trees are useful when you have large amounts of unexplored data. Partition trees are also flexible to your modeling types. The output column (Profits recode in our example) can be either continuous or nominal AND the input columns (in our example, type and #employees among others) can be either continuous or nominal. However, you should be cautious on drawing conclusions from partition trees when the data sets are small, sparse, or messy, especially as you continue to split. Methods to measure the usefulness of a partition model are supported in JMP too. To learn more about the Partition platform, go to **Help > Books > JMP Stat and Graph Guide > Chapter 37**.*

The partition tree output now also shows the probabilities for each split (see Figure 6.16). Out of the total group of companies, about 13% of them are not profitable and about 86% of them are profitable.

Now where is most of the blue? (Reminder: Blue is profitable!) It's on the left-hand branch of the splitting tree. These blue bars have been circled for you. It shows that 94% of these companies are profitable and about 5% are not profitable and come from Beverages, Drugs, Oil, and Soap. On the right-hand side, the method picked Aerospace and Computer, where 75% are profitable and 25% are not. You see more red on the right side of the tree. (Reminder: Red is unprofitable by our definition.) So we conclude that the most significant predictor of profit is the Type column, and within that column, Beverages, Drugs, Oil, and Soap are about 94% profitable. In comparison, on the right side of the tree we found Aerospace and Computer are 75% profitable and 25% unprofitable.

Based on what you just learned, what type of companies would you choose?

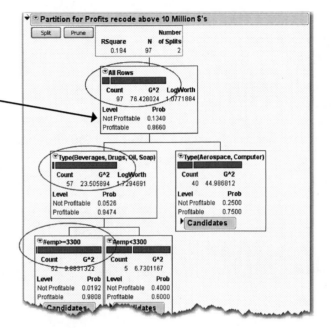

Figure 6.16

Problem Solving II

Where are the most profitable groupings after your last split (see Figure 6.17)? Hint: Follow the blue. On what side of the tree did the split happen? Can you find the combination where 98% of the companies are profitable? Hint: Look for where the number of employees in the company is greater than or equal to 3300 AND the type is Beverages, Drugs, Oil, and Soap. The best combination in the tree is circled for you.

You can now identify the most profitable combination determined so far. Let's make that easier to see.

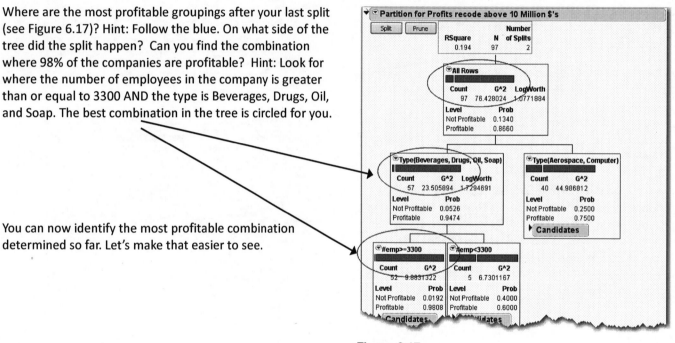

Figure 6.17

10. From the red triangle for partition, select **Leaf Report** (see Figure 6.18).

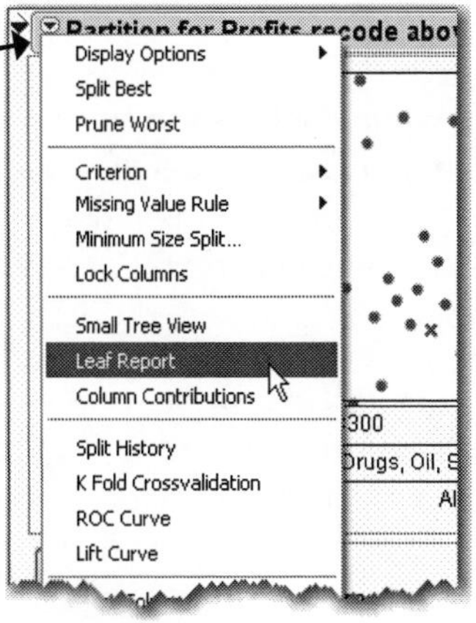

Figure 6.18

The Leaf report assembles your discoveries so far (see Figure 6.19). Looking at the Leaf report and reading the **Leaf Label** from left to right, we can see which split combinations are most or least desirable. What combination looks the most risky so far? It appears to be Beverages, Drugs, Oil, and Soap where the number of employees is less than 3300. Some 40% of companies with that combination do not achieve our definition of profitable. What group looks most profitable? It appears to be Beverages, Drugs, Oil and Soap where the number of employees is greater than or equal to 3300 with some 98% of this group as profitable. Remember this group as we'll be using it in the Using Data Filter section a bit later.

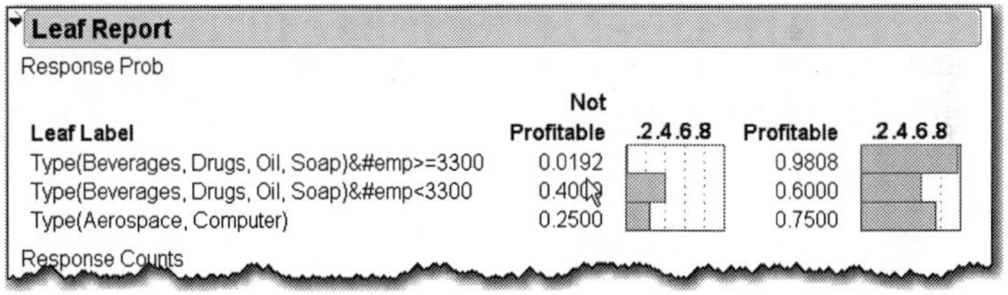
Leaf Report

Response Prob

Leaf Label	Not Profitable	.2 .4 .6 .8	Profitable	.2 .4 .6 .8
Type(Beverages, Drugs, Oil, Soap)&#emp>=3300	0.0192		0.9808	
Type(Beverages, Drugs, Oil, Soap)&#emp<3300	0.4000		0.6000	
Type(Aerospace, Computer)	0.2500		0.7500	

Response Counts

Figure 6.19

11. Click **Split** again.

Where did the next split happen? Again, look for the blue on the new split because these are profitable by our definition (see Figure 6.20). With each split, the columns in the model are sorted and ranked and the next biggest difference determines where the split will happen.

We have found another possible combination where profits are predicted on the right-hand side of the tree. The combination of Aerospace and Computer *and* Sales greater than or equal to $10,053 million has identified nine companies that are profitable. All of them are profitable in that combination. This profitable set has been circled for you (see Figure 6.20).

You can now see this last split in the updated Leaf report that appears in Figure 6.21.

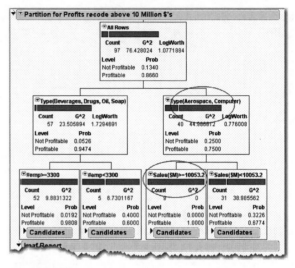

Figure 6.20

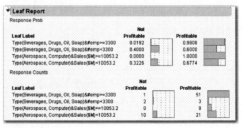

Figure 6.21

6.3 Filtering Data for Insight

Drilling down or filtering your data is an important objective in data analysis and can be accomplished in several ways with JMP. In this section we will show you three distinct ways this can be done with our example including the Tables command, the Data Filter platform, and Lasso tool. You will find that each filtering approach provides unique benefits and is suited to the context of the analysis you are performing.

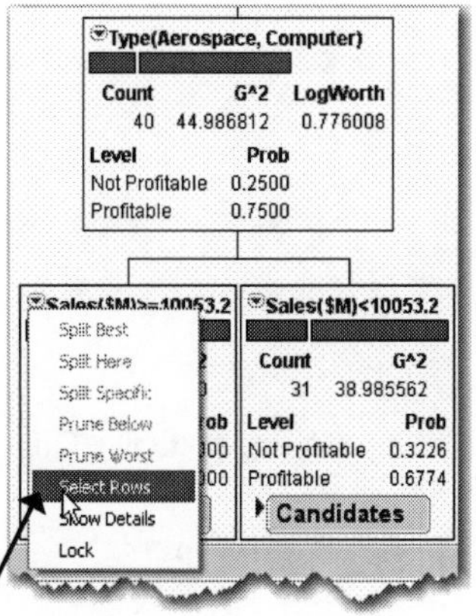

Using a Table Command to Extract a Subset

Now let's identify those nine companies:

12. Select the red triangle in the node in the tree structure where Sales($M)>=10053.2.

13. Select **Select Rows** (see Figure 6.22).

Figure 6.22

Problem Solving II

14. Select the **Tables** menu.

15. Select **Subset** (see Figure 6.23).

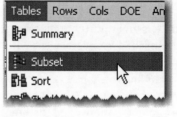

Figure 6.23

16. In the window, select **All Columns** as shown, and click **OK** (see Figure 6.24).

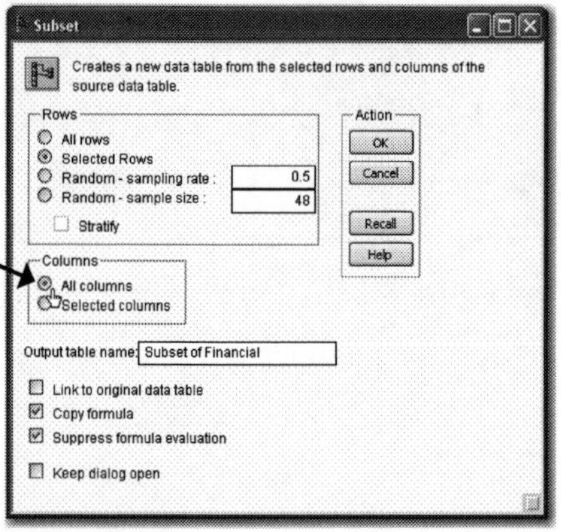

Figure 6.24

You have now identified a subset of desirable and profitable companies that are from the computer and aerospace categories and that have sales above $10,053.2 million. These companies can now be submitted for further study and consideration for the investment portfolio. If you had to report these findings, this step will have saved you from finding them in the data table. A new data table now appears (see Figure 6.25).

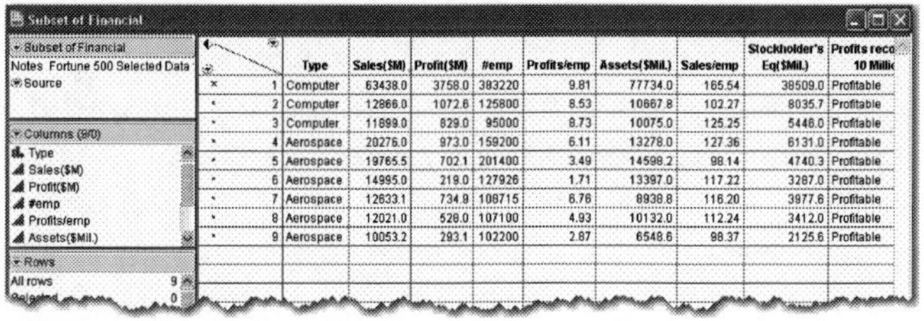

	Type	Sales($M)	Profit($M)	#emp	Profits/emp	Assets($Mil.)	Sales/emp	Stockholder's Eq($Mil.)	Profits reco 10 Milli
1	Computer	63438.0	3758.0	383220	9.81	77734.0	165.54	38509.0	Profitable
2	Computer	12866.0	1072.6	125800	8.53	10667.8	102.27	8035.7	Profitable
3	Computer	11899.0	829.0	95000	8.73	10075.0	125.25	5446.0	Profitable
4	Aerospace	20276.0	973.0	159200	6.11	13278.0	127.36	6131.0	Profitable
5	Aerospace	19765.5	702.1	201400	3.49	14598.2	98.14	4740.3	Profitable
6	Aerospace	14995.0	219.0	127926	1.71	13397.0	117.22	3287.0	Profitable
7	Aerospace	12633.1	734.9	108715	6.76	8938.8	116.20	3977.6	Profitable
8	Aerospace	12021.0	528.0	107100	4.93	10132.0	112.24	3412.0	Profitable
9	Aerospace	10053.2	293.1	102200	2.87	6548.6	98.37	2125.6	Profitable

Figure 6.25

Using Data Filter

The Data Filter command from the Rows menu is a powerful means to visualize a subset of your data interactively. The Data Filter is especially useful when you are visualizing large data tables where graphs are packed with data points and it is difficult to see meaning in them. The Data Filter allows you to easily select rows (which can be ranges or categories) within any column of interest and, in effect, hide or exclude all other rows in your data table.

We'll use the Data Filter to identify the best of the best from among the groups you already identified using Partition.

Early in the last section, recall that a promising group of companies were from the **Type** column (Soap, Oil, Beverages, and Drugs) where the number of employees is greater than or equal to 3300. We need to restrict the next analysis only to this promising subset. Here's how:

1. From the Rows menu, select **Data Filter** (see Figure 6.26).

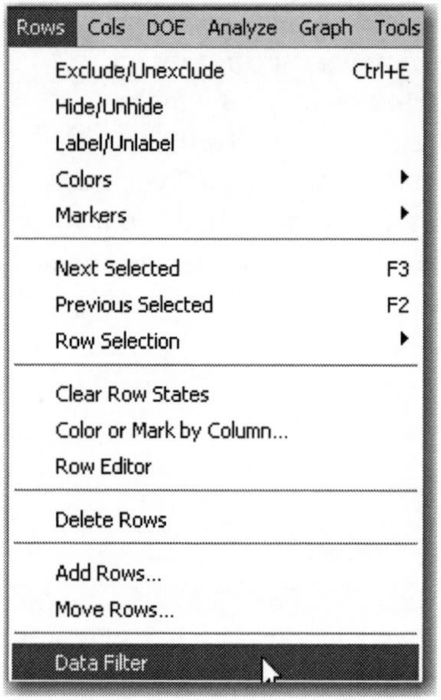

Figure 6.26

The Data Filter window appears (see Figure 6.27).

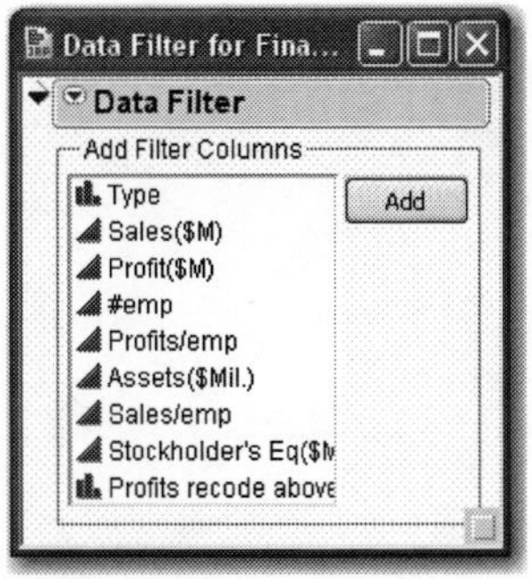

Figure 6.27

2. Select **Type** and click **Add**. A change to the Data Filter appears showing the different company types (see Figure 6.28).

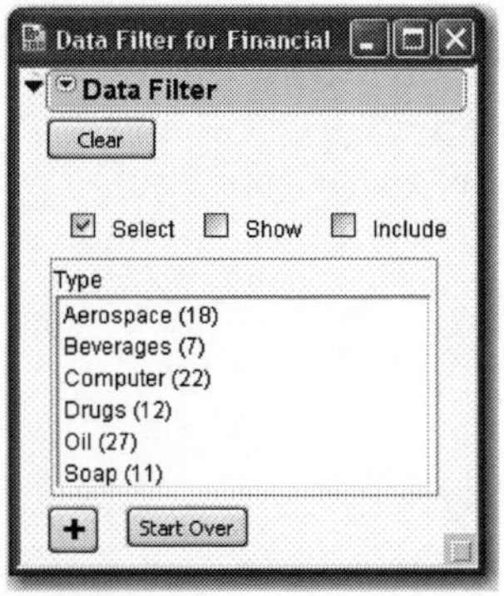

Figure 6.28

3. Now, depress the **CTRL** key and select **Soap**, **Oil**, **Beverages**, and **Drugs**. Click **+ (plus)** (results shown in Figure 6.29).

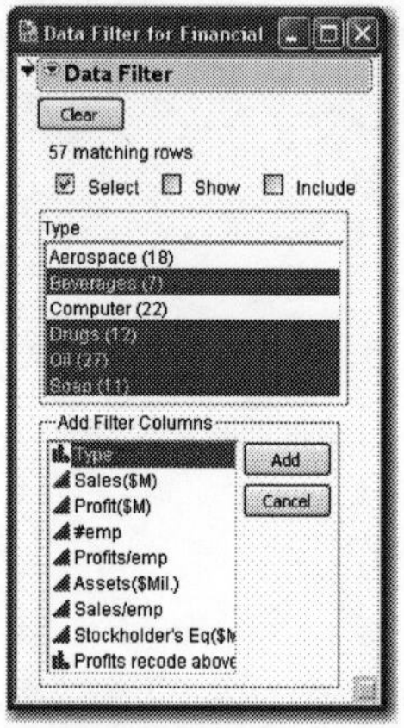

Figure 6.29

4. Select the **#emp** column and click **Add** (results with employer slider control shown circled in Figure 6.30).

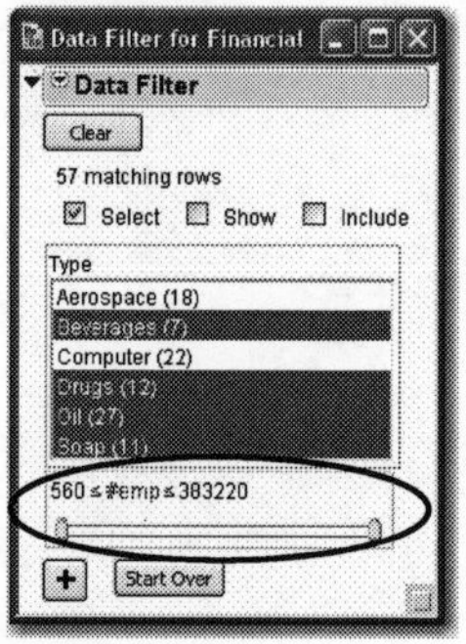

Figure 6.30

A slider appears under the range, labeled **560 <= #emp <= 383220**. Recall that the dividing point discovered in the Partition platform earlier favored companies with a number of employees greater than or equal to 3,300.

5. Click and enter **3300** where you see 560 on the left-hand side of the slider. Click the Tab key (see Figure 6.31).

6. Now select the **Show** and **Include** check boxes (see the circled area in Figure 6.32).

Figure 6.31

Note: Recall that this filter restricts the analysis to those same profitable companies identified in partition and its leaf report on page 163.

Figure 6.32

Problem Solving II

You will need to bring the data table window to the front.

7. Select **Window > Financial** (see Figure 6.33).

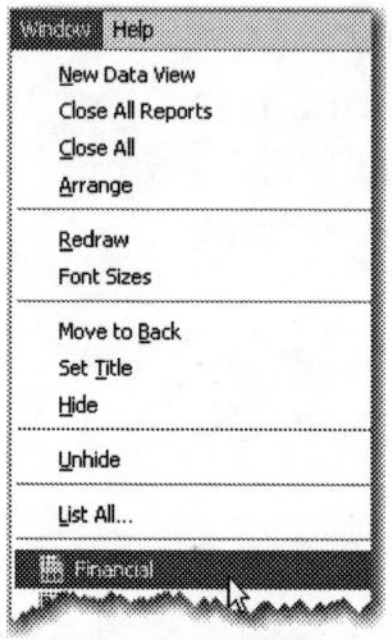

Take a look at the data table and notice what has happened. It now shows groups of rows that are selected, included for qualifying rows, and hidden and excluded for every non-qualifying group (see Figure 6.34).

The rows that are hidden (and that will not appear in graphs) feature a mask icon in the row, while rows that are excluded (not used in any analysis) feature a circle icon with a strike-through symbol.

Rows that are still included and selected (highlighted) are those that meet the criteria for success we established from our Partition example. They are now selected with the Data Filter.

Data filter is a terrific tool for exploring your data visually by allowing you to see subsets of your data in any JMP graph. As you will see in the next few pages, Data Filter will be employed to quickly subset the most desirable companies in our example.

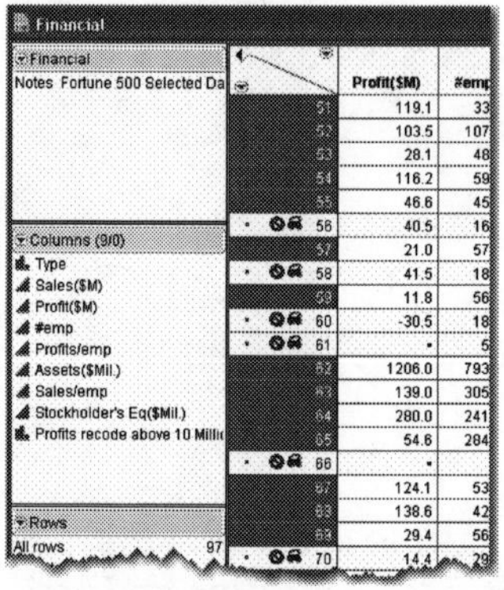

Figure 6.33

Figure 6.34

Now let's save the Data Filter for the future:

8. From the hot spot/red triangle on the Data Filter, select **Script > Save Script to Data Table** (see Figure 6.35).

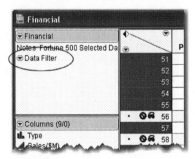

Figure 6.35

In the upper left-hand panel of the data table, a new item appears named **Data Filter**. It's circled for you (see Figure 6.36). The **Data Filter** item has stored the steps needed to reproduce the subset you selected in a JMP script.

Figure 6.36

Let's rename it something more meaningful so we can remember what it does:

9. Double-click on **Data Filter** and name it **SoapOilBevDrug, Emp >= 3300** in the window that appears (see Figure 6.37).

10. Click **OK**.

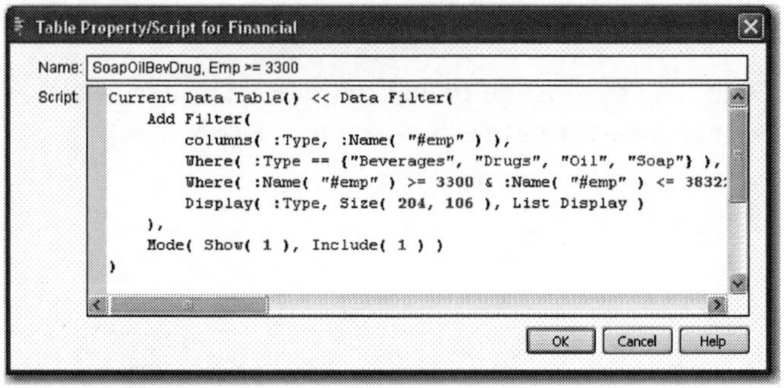

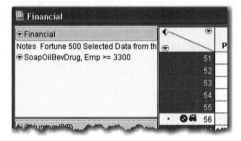

Figure 6.37

The Data Filter script is now renamed SoapOilBevDrug, Emp>= 3300 in the upper left-hand panel of the data table (see Figure 6.38).

Now we can reuse the filter if we want to apply this criterion to the data. It's easy with the new table property that applies the filter.

Note: The script also will act upon any new data you add or merge into the data table.

Figure 6.38

You have now filtered your data to only the best performers, so our task is to select only the very best from this group* for our portfolio. There are many methods to finding high performers from among the groups we selected. We will go back and use the Fit Y by X platform (that we introduced in the previous chapter) along with the Lasso tool to visually select individual points. We find that using Lasso in this context easily enables visual selection of points that meet our performance requirements. The goal is to select the most profitable companies by industry type. In this example we'll use the continuous representation of profits, **Profits($M),** as it allows us to distinguish with more precision among this group of the high performers.

11. Select **Analyze > Fit Y by X** (see Figure 6.39).

12. In the resulting window, **select Profits($M)** as the **Y, Response** and **select Type** for the **X, Factor** (see Figure 6.40).

13. Click **OK**.

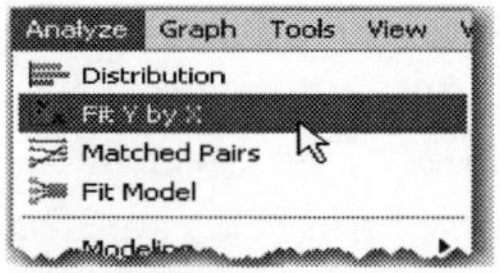

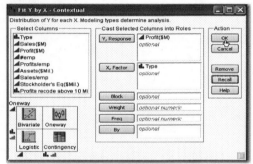

Figure 6.39

Figure 6.40

*Note: Recall that we also identified profitable companies from the Aerospace and Computer types in the partition analysis and we used the subset command to extract those. Because these companies also had the requirement of having sales over $10B, we could have used a conditioning statement (such as an "and" or "or" statement) in the Data Filter or merged that group with the one illustrated above to create one data table. To simplify this illustration we've omitted these steps. To learn about how to merge data tables, see **Help > JMP User Guide > Chapter 7 Reshaping Data** and the topic, Concatenate.*

The oneway graph (see Figure 6.41) shows profits from one of the best performing groups you identified in your partition analysis and selected using the Data Filter. All other companies have been excluded and hidden from this analysis. It is from these included companies that we want to extract the very best of these performers.

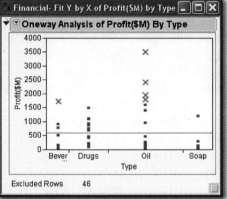

Figure 6.41

14. From the hot spot/red triangle, select **Means/Anova** (see Figure 6.42).

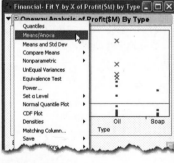

Figure 6.42

The result appears in Figure 6.43.

Notice the means diamonds for each of the groups that we learned about in the previous chapter. The height of the diamond corresponds to a 95% confidence interval. Use the question mark tool from the **Tools** menu and click on a diamond to review confidence intervals.

Individual data points above or below the diamonds fall outside the confidence interval. We are interested in those above the diamonds because they are exceptional performers in terms of profit. Can you think of a way of selecting them?

Using Lasso to Select Individual Points

Here's a new way:

15. From the toolbar, select the **Lasso tool** (see Figure 6.44).

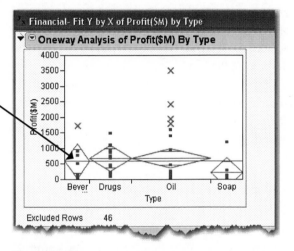

Figure 6.43

Figure 6.44

16. Now move the tool to the **Oneway** result. Left-click and draw around the points above the diamonds (see Figure 6.45), making sure that you fully encircle the points before releasing the button.

There should be nine selected rows in your table. Now we just need to extract these very best historical performers from the rest.

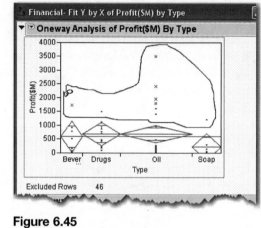

Figure 6.45

17. From the **Tables** menu, select **Subset** (see Figure 6.46).

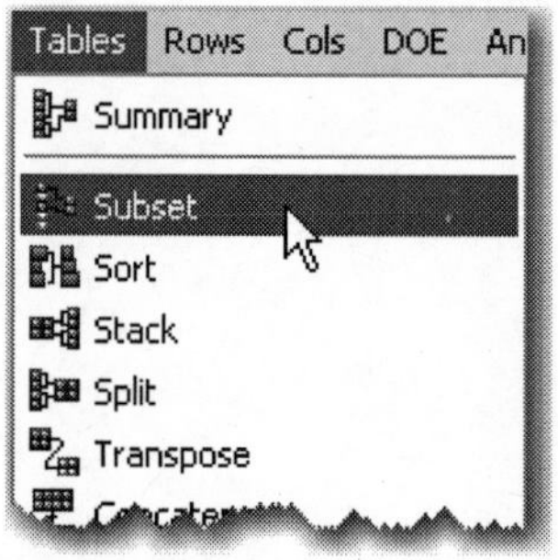

Figure 6.46

18. In the resulting window, select **Selected Rows** and **All Columns** (see Figure 6.47).

19. Click **OK**.

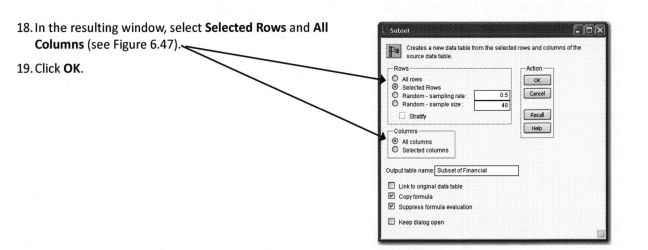

Figure 6.47

The resulting data table (see Figure 6.48) contains nine companies that have been sifted down from a group of 97 that represent the most profitable. The most profitable companies are mostly oil companies, though soap, drugs, and beverage companies each made a showing among the most profitable. We now have some potential portfolio selections for further study or investment.

	Type	Sales($M)	Profit($M)	#emp	Profit
1	Drugs	6698.4	1495.4	34400	
2	Oil	86656.0	3510.0	104000	
3	Oil	50976.0	1809.0	67900	
4	Oil	32416.0	2413.0	37067	
5	Oil	24214.0	1610.0	53648	
6	Oil	21703.0	1405.0	31338	
7	Oil	15905.0	1953.0	26600	
8	Soap	21689.0	1206.0	79300	
9	Beverages	9171.8	1723.8	20960	

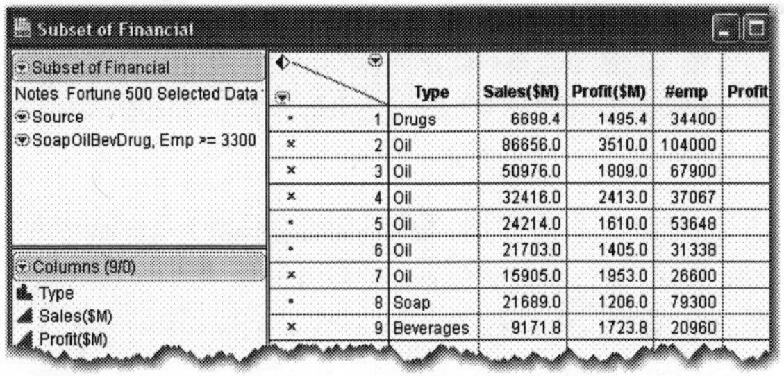

Figure 6.48

6.4 Model Fitting, Visualization, and What If Analysis

In this section we will extend many of the JMP skills and concepts you've learned so far into the insight you'll need to articulate your data effectively. We will now combine our models with the Data Filter to better understand the dynamics of our data. We will also introduce the Prediction Profiler for real time visual what-if analysis.

Consider for a moment a thought experiment. Some old radios still have an analog tuner dial on them. Imagine you turn on the radio and hear noisy static. You turn the tuner dial slowly through the frequencies until you hear a transmission. Sometimes you might tune past the station, but you slowly tune back to obtain the optimal signal. Your ear heard the difference between the signal and the background noise. The Profiler and Data Filter (as described in the previous section) can be used as just such a tuner, but instead of using your ears, you use your eyes. The next section shows you how.

Let's test the idea that a positive relationship exists between profits and the number of employees in all of our companies. To do this, we need to use two things you already learned: Fit Y by X and the Data Filter.

1. Select **Rows > Clear Row States** (see Figure 6.49).

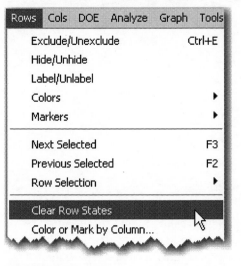

Figure 6.49

This clears the row selections, the markers, and the hidden and excluded rows from the last section.

Model Fitting

Now let's build a simple model of the relationship between profits and number of employees:

2. Select **Analyze > Fit Y by X**. Select **Profits($M)** as the **Y, Response** role and **#emp** as the **X, Factor** role (see Figure 6.50).

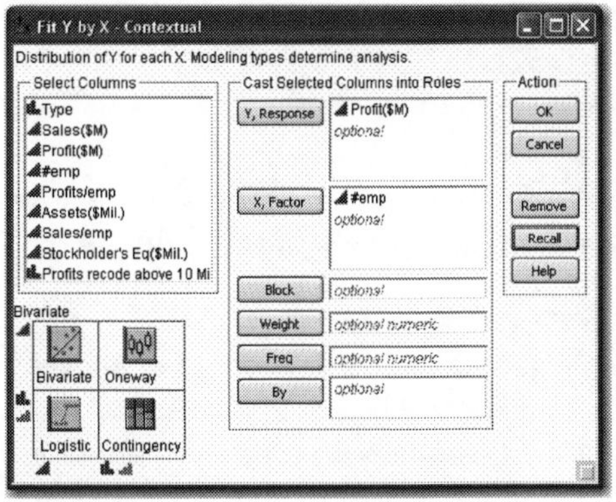

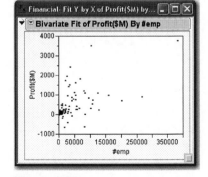

Figure 6.50

3. Click **OK**. A bivariate fit appears (see Figure 6.51).

Figure 6.51

The scatter of the points might seem to indicate that as the number of employees increases, the profits also increase. Let's fit a line through the points to help us understand the relationship better.

Problem Solving II

4. From the hot spot/red triangle for the bivariate fit, select **Fit Line** (see Figure 6.52).

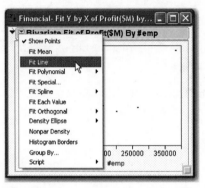

A red line appears in the data that extends generally from the lower left of the graph to the upper right of the graph (see Figure 6.53).

This line* shows that, in general, as the number of employees increases, so do the profits. Notice how sparse the data is for companies above 150,000 employees. Because there are few companies or rows of data above 150,000 employees, it might suggest that the relationship is weaker for very large companies.

Figure 6.52

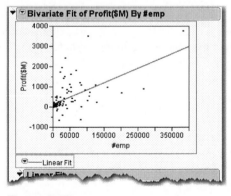

This is a simple or linear least squares regression fit, which fits a straight line through the points in a manner that balances the differences between those points above and below the line.

Figure 6.53

Let's see if graphics can help us see this relationship:

5. Select the red triangle item next to the line labeled **Linear Fit** and select **Confid Shaded Fit** (see Figure 6.54). A pink shaded boundary around the line appears (see Figure 6.55).

The curves appearing on the graph are 95% confidence curves. The shaded area around the line gets wider where data is sparser. This is a visual indication that where data is sparse, the estimates get wider. The converse is also true: where data is denser, the confidence curves are narrower. Thus, we can be more confident that the line better represents the relationship where the confidence curves are narrower than where they are wider. This might indicate that the relationship between profits and number of employees is strongest up to about 150,000 employees because the confidence curves start to widen there.

Notice, too, that a lot of the data points are pretty far away from the line. That's curious. It appears especially true for companies with more than 50,000 employees. Maybe even more than one pattern is present. How might we explore this?

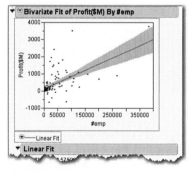

Figure 6.54

Figure 6.55

Problem Solving II

6. Select **Rows > Data Filter** (see Figure 6.56).

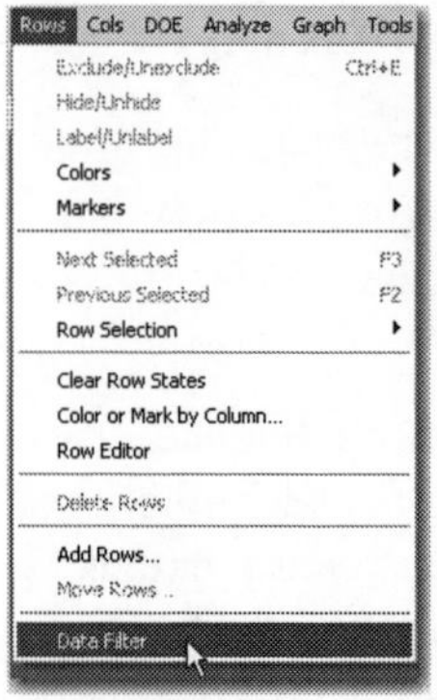

Figure 6.56

7. In the Data Filter window, select the **Type** column (see Figure 6.57) and click **Add.**

Figure 6.57

8. Select the **Show** and **Include** check boxes (see Figure 6.58).

Figure 6.58

9. Now return to your Fit Y by X results. From the red triangle, select **Script > Automatic Recalc** (see Figure 6.59).

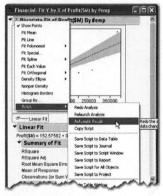

Figure 6.59

Adding **Automatic Recalc** creates a live link via the Data Filter between the rows in the data table and the Fit Y by X graph window.

Now, you should see two floating windows (see Figure 6.60).

The goal now is to see if the positive relationship between profits and number of employees holds across all company types. The Data Filter acts like a tuner as it toggles through the company types. Here's how:

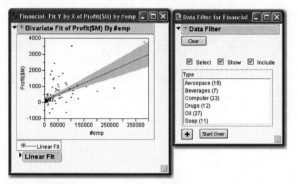

Figure 6.60

10. From the hot spot/red triangle of the Data Filter, select **Animation** (see Figure 6.61). Some animation controls appear in the Data Filter window that look like the controls on a DVD player.

Figure 6.61

11. Click on  (the **step forward** control) (see Figure 6.62).

Each time you click the step forward control, you are filtering the Fit Y by X analysis to just one of the type groups you see highlighted in the Data Filter.

Figure 6.62

Now click sequentially through the company types and notice what happens to the graph each time you click. Toggle through several times. Can you see the one type that is different from the rest? Figure 6.63 includes the series of graphs you should be seeing as you toggle through each company type.

Which graph is most different from the rest? Which graph has a fit line that is almost flat? You might have noticed that Beverages shows a line that is much flatter than the others, which indicates that number of employees or size of company has less influence on the profitability. Notice that in general the other company types show a strong correlation between profits and number of employees.

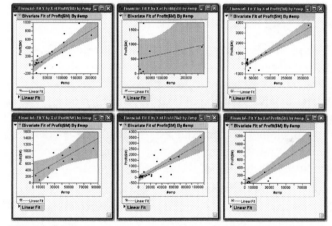

Figure 6.63

---

*See Appendix B for more information.

Notice also that Beverages is the sparsest fit, with only seven data points widely dispersed so the confidence bands also flare out widely. At least for the beverage company type, we can conclude that the greater number of employees does not predict greater profits in this sparse sample.

12. Select **Rows > Clear Row States** for next section.

Conducting What If Analysis

The Prediction Profiler is a different type of graphical tuner for your data and provides a clear picture of your model. The Prediction Profiler is an interactive graph that produces estimates of your Y column of interest (profits, in our example) subject to your predictors, or X, columns, (such as number of employees from the last example). The interactive feature allows you to drag and change settings of any column to see the estimated effect on the other columns.

The advantage of the Prediction Profiler is that it lets you try what if scenarios dynamically and get immediate estimates on any column of interest. Very cool!

1. To create a profiler, you first need to create a model to describe the relationships in your data. Select **Analyze > Fit Model** (see Figure 6.64).

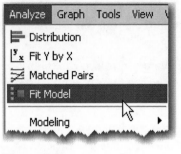

Figure 6.64

Caution: The Prediction Profiler appears in the context of a multiple regression model here. The technique is introduced here to enable quick exploration of your model. It is an advanced analytical method. You should familiarize yourself with the underlying analysis concepts by using the question mark tool to call up the relevant chapters in the documentation.

2. In the Fit Model window, select **Profits($M)** as the Y column and then select **Sales($M)**, **#emp**, **Assets($Mil.)**, and **Stockholders Eq($Mil.)**. Click **Add** to place them in the Construct Model Effects window (see Figure 6.65).

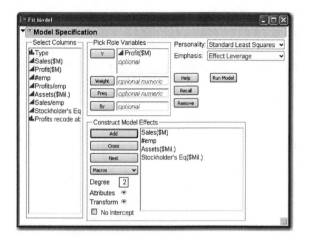

Figure 6.65

3. Select the pop-down menu and change it from **Effect Leverage** to **Effect Screening** as shown (see Figure 6.66).

Figure 6.66

Now we are ready to run the model to test the relationships between profits and the other columns we've selected. This approach using the Fit Model platform employs a method called *multiple regression*. We'll exclude the type of company for now from the model since we will be using it later to investigate differences.

4. Click **Run Model**. A report window appears (see Figure 6.67).

The Actual by Predicted Plot indicates there is a strong, positive correlation between profits and the columns you picked because of the 45-degree angle of the red fit line relative to the blue mean line. There are other technical indicators of the quality of the relationships in the other report items. Use the question mark tool (**?**) to investigate the report items and learn more about the results.

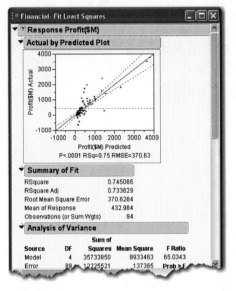

Figure 6.67

5. Scroll down the window until you see the Prediction Profiler (see Figure 6.68).

Prediction Profiler

The Prediction Profiler enables you to see visual relationships between profits (Y column) and sales, number of employees, assets, and stockholder equity (X columns), as depicted by the fitted solid black lines (see Figure 6.68).

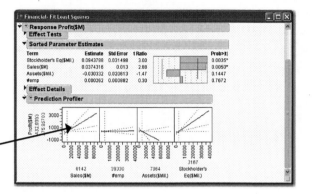

Figure 6.68

Note: Each graph is bisected by a vertical red dotted line. These red dotted lines are interactive and provide us with superhero-like powers.

6. Drag the vertical dotted red line for sales from its current set point near 0 to a new set point near 40000 (see Figure 6.69).

As you drag the slider for sales, notice that the estimate for profits increases to around 1625. (You may not hit it exactly. This is okay.)

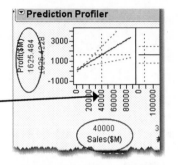

Figure 6.69

Each of the vertical and horizontal red dotted lines in the Prediction Profiler are interactive and work in this manner. Look at the angle of the fitted line in the plots within the Prediction Profiler (see Figure 6.70).

With sales, for example, this is an indication that our column of interest (profits) changes a lot when we move the red dotted vertical line associated with **Sales($M)**. Thus, the steeper the line, the greater effect a change has on our Y column of interest. Lines that are flat, or nearly so (such as employee), mean that changes in these columns have little impact on our Y column (profits). With the Profiler, you can conduct dynamic, real-time what-if analysis.

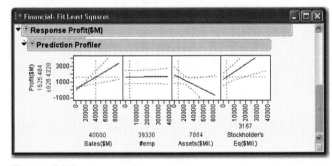

Figure 6.70

We previously learned that there was variation in profits among some company types. However, the model we just created did not include the Type column. So let's explore how each type of company might impact profits in this model. We again use the Data Filter to accomplish this:

7. Select **Rows > Data Filter** (see Figure 6.71).

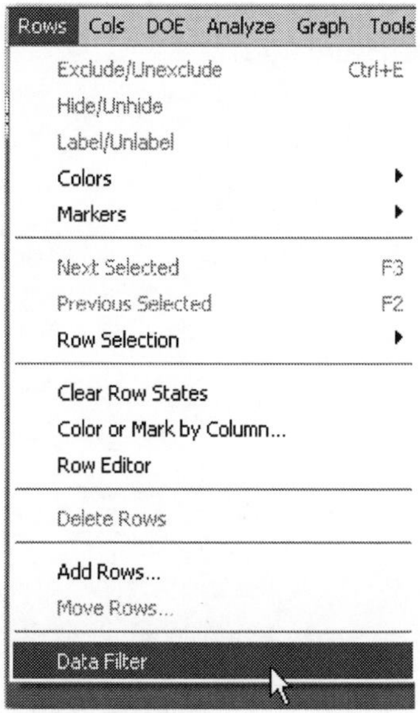

Figure 6.71

8. In the Data Filter window, select **Type** and click **Add**. Select **Show** and **Include** (see Figure 6.72).

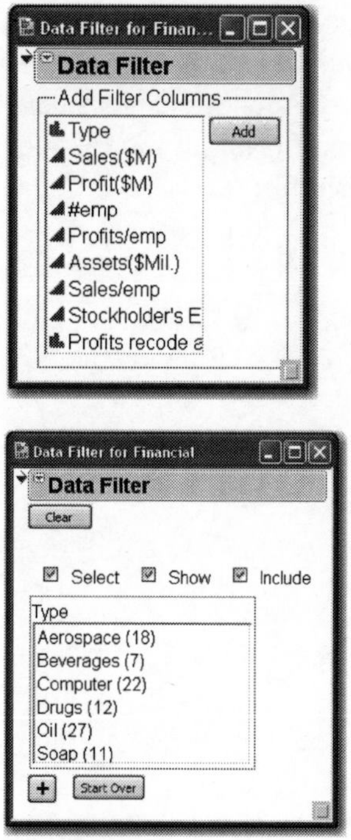

Figure 6.72

We need the report window to respond to the Data Filter dynamically using automatic recalculation. To do this, follow these steps:

9. From the hot spot/red triangle in the report window that contains the Prediction Profiler, select **Script > Automatic Recalc**. The Prediction Profiler now responds to changes you make in the Data Filter.

10. In the Data Filter window, click on the hot spot/red triangle and select **Animation** (see Figure 6.73).

11. Click on (the step forward step control) (see Figure 6.74).

12. Now click the step button again (which toggles through the company types). Watch what happens to the Prediction Profiler with each click as it steps through the company types.

Figure 6.73

Figure 6.74

Have you noticed that some of the fitted lines in the profiler flip directions as the company type changes? How would you interpret these changes? We have summarized the first few of the profilers you see as you click through the company types. We will also provide a brief interpretation of each profile. Within each type, we can still drag the red dotted lines to investigate the sensitivity analysis or relationships within the model for that company type.

Aerospace

As sales increase, profits decrease (see Figure 6.75). As the number of employees increases, profits increase. As assets increase, profits decrease. As stockholder's equity increases, profits increase.

Beverages

The confidence bands are so far away from the fit that we must be very cautious about trusting the profiler for Beverages (see Figure 6.76).

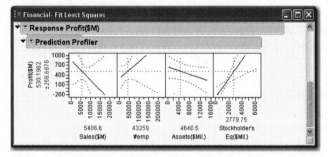

Figure 6.75

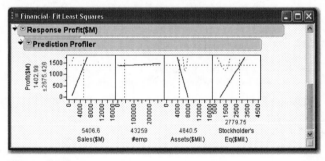

Figure 6.76

Computer

As sales increase, profits increase (see Figure 6.77). As the number of employees increases, profits decrease slightly. As assets increase, profits decrease steeply. As stockholder's equity increases, profits increase.

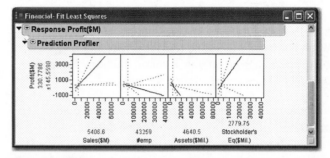

Figure 6.77

Soap

As sales increase, profits increase (see Figure 6.78). As the number of employees increases, profits stay the same. As assets increase, profits decrease. As stockholder's equity increases, profits increase.

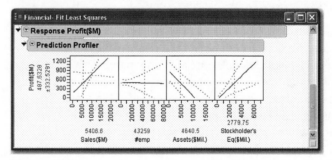

Figure 6.78

Now it's your turn. On your computer screen, try reading the Prediction Profiler for Oil using what you have learned.

6.5 Summary

Most real-world problems involve multiple columns. The Partition platform is a flexible tool for solving many types of problems that involve multiple columns and rapidly identifies the key relationships in the data.

Interactive tools like Data Filter and Lasso enable quick identification and extraction of interesting data subsets. The Data Filter also provides an exploration method using animation that lets you tune in to just the slice of the data that best supports an inference or hunch.

The Prediction Profiler using the Fit Model platform offers a powerful way to understand the relationships among columns in your models. The ability to manipulate values (using the red dotted lines) within a column and immediately see their effect on other columns provides the means to conduct visual what if analyses. With the Data Filter, you can drill down to subcategories or ranges of columns to discover those nuggets of insight that are often hidden at first glance.

This chapter and the last have presented an approach to problem solving that is unique to JMP. The approach underscores the progressive nature of discovery that tends to build from simple descriptions of one column of data to complex relationships among many columns. This problem-solving process leads to a better understanding of your data and, in turn, to the insights and answers you seek. This process might not only go in one direction from simple to complex. As discoveries are made with the more advanced multi-column tools, confirmation and further analysis can be made with the simpler ones, Distribution and Fit Y by X, from where our journey began.

Chapter 7 • Sharing Graphs

This chapter focuses on common ways you can customize and share JMP graphs with others whether you are presenting those findings yourself, placing them in a document, or sending a file. We also provide some advice for effectively doing so.

One of the most important things a JMP user can do is communicate results (data, statistics, and graphs) clearly and accurately. Most people recognize that it can be easy to manipulate the results or create a graph that doesn't tell the whole story. The first section of this chapter is devoted to principles of effective communication with graphs.

Moving *static* graphs into other applications, such as Microsoft Word and PowerPoint, is a straightforward operation in JMP. Because this is a very common need among JMP users, we also illustrate some options that allow you to edit the results in other applications.

Some of JMP's new *animated* graphing features now provide the ability to move the graphs to the Web or into other applications while retaining many of the animation capabilities. Using this feature is a little more involved, but we walk you through the necessary steps to accomplish this in Section 7.5.

If you want to annotate a graph with comments or change colors or other settings before moving it into another document, JMP provides several convenient tools introduced in Section 7.2. Further, you might want to annotate and put several graphs or results together into a single image to provide a more complete story of your results. The JMP Layout tool provides this functionality; we cover its essentials in Section 7.6.

Sharing Graphs

7.1 Presenting Graphs Effectively

Presenting statistical graphics in a report or presentation is a common task for many JMP users. As such, it is critical that you do this well. By *well*, we mean that you rely upon the data to tell the story, be succinct, and, most importantly, present an accurate and valid interpretation (visual, numeric, verbal). It helps to keep the following in mind when presenting statistical graphics:

1. Understand your data and its limitations. Time spent with your data before you present it prepares you for the questions that may emerge.

2. Focus on the important factors. What columns are core to the problem or question at hand?*

3. Provide the simplest expression of the data to convey the most complete description of it.

4. Present graphs accurately. Do not reduce ranges in graphs to magnify changes or increase ranges to hide trends.

5. Let graphs speak for themselves and avoid using unnecessary background colors or dimensions that distract from the core information you are conveying. Make sure the graph can be seen by your audience.

An important feature of JMP is that it helps you adhere to these principles and commonly accepted standards. JMP can provide many different types of appropriate graphs of your data, so how do you know which is the most effective way to communicate your information? You will find that using common sense goes hand-in-hand with effective presentation of data and graphical integrity.

In recent years, there have been a number of new books on the subject of presenting quantitative graphs and visualizing data effectively. Perhaps the best-known of these are books by Edward Tufte, whose beautifully crafted books have won wide acclaim and illustrate the core principles of presenting quantitative information effectively. These books and those by Stephen Few (see the Bibliography) are highly recommended if you use graphs extensively.

How do you determine which variables or columns are most important? The methods described in Chapter 6 help you identify important columns of interest. For example, the Partition platform (Section 6.1) can help you find out which columns are the key factors affecting another variable.

Sharing Graphs

7.2 Customizing Graphs for Presentation

There are many features in JMP that provide you with the ability to customize graphs for presentation. Sometimes you'd like to annotate a graph by pointing out a key attribute of the graph. Or, you might want to use color or markers to enhance other columns and attributes, as described in Section 2.5. This section introduces the common tools needed to prepare your graphs and results before moving them into another application or document.

Most of JMP's options appear by simply right-clicking on the area (for example, the graph or axis) you'd like to customize. When customizing your graphs, be sure to make changes within JMP before pasting them into another application. The first step, however, is to look at a powerful set of tools called the JMP Toolbar that we begin with in this section.

Example 7.1 SAT by Year

We will be using the SATByYear.jmp data table to illustrate the concepts in this chapter. SAT by Year is data representing SAT (Scholastic Aptitude Test) and ACT (Achievement Test) test performance by college-bound students in the United States covering the years 1992 to 2004. The tests serve as an important admission metric for colleges and universities.

The data consists of 17 columns and 408 rows of data and includes the following columns:

State: Students' state of residence
Expenditures (1997): State expenditures per pupil
Student/Faculty ratio (1997): Ratio of students to faculty
Salary (1997): Mean salary of teachers (in thousands)
%Taking (2004): Percent taking the SAT exam in 2004
Population: Population of the state
% Taking (1997): Percent of students taking the SAT in 1997
ACT Score (2004): Mean test scores by state in 2004
ACT % Taking (2004): Percent taking ACT in 2004
ACT Score (1997): Mean test scores by state in 1997
ACT %Taking (1997): Percent taking ACT in 1997
Year: Year of the tests
SAT Verbal: Mean SAT Verbal test score by state
SAT Math: Mean SAT Math test score by state
Region: Region in which state resides

You can access this data at **Help > Sample Data > Open the Sample Data Directory > SATByYear.jmp.**

Using the JMP Toolbar

Among other things, the JMP Toolbar contains several tools that help you copy, annotate, and share JMP graphs. These tools are useful in the JMP Layout tool (see Section 7.6), where you might construct a report. Many of these tools have familiar functions:

The **Selection** tool allows you to select graphs or reports (or portions of them) to be copied and pasted into Word, PowerPoint, or other applications.

Within the Toolbar, the following functions allow you to add comments or highlight areas of your graphs:

-**Annotate** allows you to add a textbox or comment in your graphs.

-**Lines** allow you to draw lines on your graphs. For example, an arrow drawn from a textbox to a point on a graph.

-**Polygon** and **Simple Shape** allow you to draw shapes in your graphs.

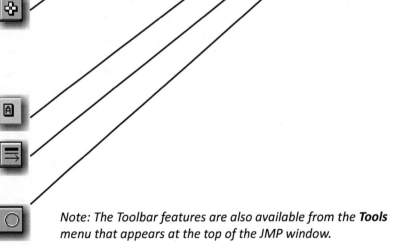

*Note: The Toolbar features are also available from the **Tools** menu that appears at the top of the JMP window.*

Figure 7.1 is an example of what these tools can do for you. You can summarize key aspects of your data with the Annotate tool and create pointers using the Lines tool to call out this information. Interpretation of this kind can provide valuable direction when you are not there to guide the report recipient. To generate these graphs in the **SATByYear** data table, select **Analyze > Distribution > Student/Faculty Ratio** and **ACT Score, Y, Columns > OK**.

To create these features in JMP, do the following:

1. Select the **Annotate tool** from the JMP Toolbar. Click on your report where you want to add an annotation. Type your comment and then left-click outside the annotation box. You can then move or resize the box as you would any textbox or change the color of the background by right-clicking on the box.

2. To add lines or pointers, select the **Lines** icon from the Toolbar and draw the line. Once the line has been drawn, right-click to access the options for pointers, thickness, style, and color.

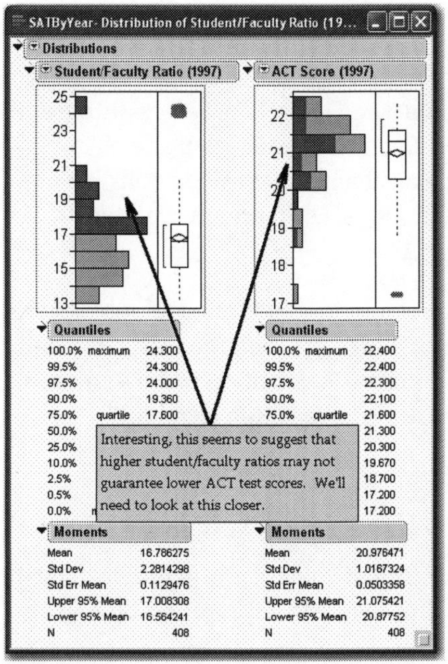

Figure 7.1

When you'd like to annotate more than one graph window and show these together, use the Layout tool, which is discussed in Section 7.6.

Sharing Graphs

Using Color

Changing the colors of your basic graphs is simple. Just right-click within the graph area (see Figure 7.2). Select **Histogram Color** and select a color. If you want your graphs to always use the same color, you can change the JMP defaults by selecting **Save Color Preference**. For more information on preferences, see Chapter 1.

Coloring a graph in this manner does not affect graphs where individual points are color-coded by a column within the data table. (See Section 2.5 for more information on how to use color to denote a category or range within a graph.)

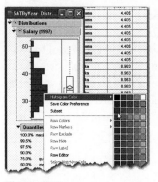

Figure 7.2

Background Color

You can change the background color of any graph, by right-clicking in a graph area and selecting **Background Color** (see Figure 7.3).

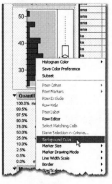

Figure 7.3

Horizontal Layout

Within the Distribution platform, the default display places the histograms in a vertical (side-by-side) form. Clicking on any bin (or bar) within one histogram highlights those values in the remaining histograms. This is especially useful when you are exploring relationships among one or more columns and don't want to scroll down.

Sometimes, however, you'd like to see the histogram in the more traditional horizontal format. Rotate the display by simply selecting **Display Options** under the red triangle after you've created a histogram and then selecting **Horizontal Layout** (see Figures 7.4 and 7.5).

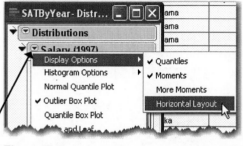

Figure 7.4

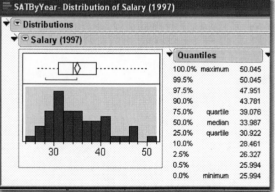

Figure 7.5

Note: If you want to stack multiple histograms horizontally, one on top of the other, you can use the Stack command under the Distributions red triangle (immediately above Salary's red triangle illustrated in this example).

Sharing Graphs

Axis

JMP allows you to easily change the axis of any graph by simply grabbing the axis and moving it. Alternatively, you can right-click on the axis to access the following menu (see Figure 7.6).

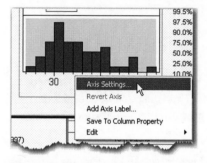

Figure 7.6

By selecting **Axis settings**, you access the Y Axis Specification window (see Figure 7.7). In it, you can customize the range, increment, and other axis options.

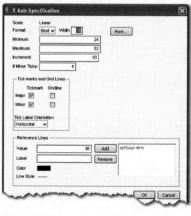

Figure 7.7

7.3 Placing Graphs into PowerPoint or Word

Copying a fixed JMP graph into Word or PowerPoint is simple. You just need to use the Selection tool to select what you want and then copy and paste the graph into your document.

Let's illustrate this with the following example. Say you want to move the graph in Figure 7.8 from JMP into PowerPoint.

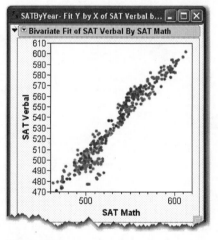

Figure 7.8

1. Select the **Selection tool** (see Figure 7.9).

Figure 7.9

2. Left-click on the upper left corner of the graph, as shown, to capture the entire picture (see Figure 7.10).

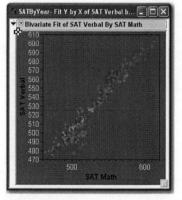

Figure 7.10

Alternatively, you might want to copy only selected elements of the graph, in which case you left-click and drag the cursor to select the elements you'd like to copy.

3. Right-click and select **Copy** (see Figure 7.11).

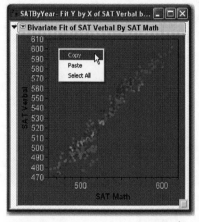

Figure 7.11

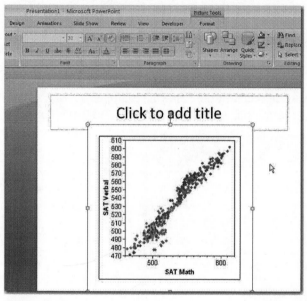

Figure 7.12

4. Open the desired application (PowerPoint in this case), open the presentation and create a slide with the desired layout, right-click, and select **Paste** (see Figure 7.12).

Note: When using JMP's default paste function, the object can be re-sized in the new application, but the contents are fixed.

Sharing Graphs

Special Paste Functions

JMP also provides additional paste formats and functions. Sometimes you might want to edit the results or graph or you need to provide the graph in a specific format. JMP has some options for you:

- Use **Paste Special**. This allows you to paste a .bmp or metafile (see Figures 7.13 and 7.14).

Figure 7.13

> Metafile: a vector art graphics format that can be edited in detail by Word and PowerPoint on Windows.

Figure 7.14

- If you want to change the format of all the graphs you want to paste, change your preferences to reflect this:

  1. Select **File > Preferences**.

  2. Select **Windows Specific** (or **MAC OS**).

  3. Under Graphic format for RTF files, select **Metafile** from the drop-down menu (see Figure 7.15).

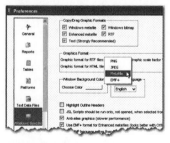

Figure 7.15

Sharing Graphs

7.4 Creating and Sharing Animated Graphs

Animated, or motion-enabled, graphs are used to express trends among column/variable relationships. JMP provides the means to communicate a clear impression from a complex model and, with the Bubble Plot and Profiler, you can export these to other documents. Unlike other graphs that are static or fixed, this new class of graphs brings an entirely new dimension to your data by allowing you to animate changes in data relationships over time, range, or category—very much like the animated cartoons we all watched growing up, except now you can interact with them.

In JMP 7, users could only paste a static snapshot of a motion-enabled graph into another application. In JMP 8, however, JMP's motion-enabled graphs (specifically the Bubble Plot and the Profiler) can be saved as a Flash file or moved into other applications using Adobe® Flash®, where they can be expressed in motion. What's nice about this new feature is that the Flash file that is created still contains many of the functions that are native to these graphs inside JMP—very cool!

In this section, we show you how to create the Flash file for these graphics and, in the following section, how to access

and use these files in a PowerPoint presentation. We use the Bubble Plot to illustrate this section with time series data (steps for the Profiler are essentially the same; see Chapter 3.) Time series data is good to illustrate with motion because it inherently reflects change over time.

Creating Exportable Motion-Enabled Graphs

The Bubble Plot is found in the **Graph** menu. The window is shown in Figure 7.16. As you see, the Bubble Plot allows you to visualize up to six variables at once, including points, x- and y-axes, time, size of bubble, color gradient, and By. JMP allows you to also select a second subgroup for your points. For example, if you have sales data that contains both individual territories and regional information, you might want to toggle between **Region** and **Territory**. This feature allows you to see the motion-enabled graph for the region and to split that region into its territories on the fly to see their individual performance.

Sharing Graphs

1. Open your data file and select **Graph > Bubble Plot**. Select **columns** to match desired roles (described in the following box). Only Y and X are required, but you will want to do more with this platform (see Figure 7.16).

> *Using our chapter example SAT by Year, assign these columns to their corresponding roles:*
>
> | Column | | Role |
> |---|---|---|
> | *SAT Verbal* | > | *Y* |
> | *SAT Math* | > | *X* |
> | *Region* | > | *ID* |
> | *State* | > | *ID* |
> | *Year* | > | *Time* |
> | *% Taking (1997)* | > | *Sizes* |
> | *Region* | > | *Coloring* |

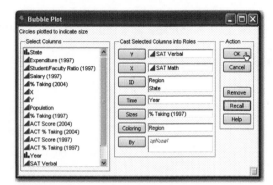

Figure 7.16

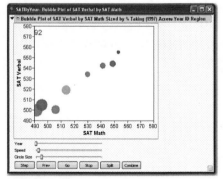

Figure 7.17

2. After selecting **OK** (see Figure 7.17), locate the red triangle in the top left corner.

Sharing Graphs

3. Under the red triangle, there is a menu option, **Save as Flash (SWF)** (see Figure 7.18). Selecting this option saves the file (as a .swf file), along with a companion .htm file. Name your file and select **Save**.

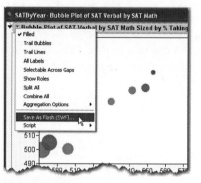

Figure 7.18

4. It takes a few seconds, but you will see that a Flash version of that Bubble Plot now automatically appears on your desktop (in your browser). You can post it to a Web site (see Figure 7.19).

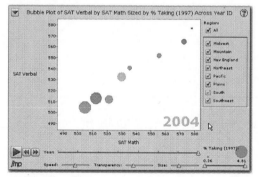

Figure 7.19

Sharing Graphs

7.5 Placing Animated Graphs into PowerPoint

Before moving these graphs into PowerPoint, first save your graph as a Flash file (as described in the previous section). Once you have done this, you will point to the .swf file from inside PowerPoint. This process is a little more involved, but we walk you through the steps here:

1. Save your Shockwave Flash (".swf") file. Note the file name and location on your computer.

2. Create a PowerPoint slide with a layout format that provides ample room to display the animations, such as Title and Content (see Figure 7.20).

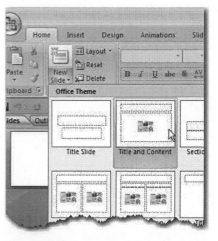

Figure 7.20

3. Once you have created the PowerPoint slide, select the **Developer** tab (see Figure 7.21).

Note: We are illustrating this example with PowerPoint 2007. If you use other versions of PowerPoint the steps may differ. If needed, consult PowerPoint Help or **www.jmp.com/support**.

Figure 7.21

4. Select (the More Controls icon) to view more controls (see Figure 7.22).

Figure 7.22

5. Select **Shockwave Flash Object**, and click **OK** (see Figure 7.23).

Figure 7.23

Sharing Graphs

6. On your PowerPoint slide, adjust the box to the right size to accommodate the Flash-animated graph. For this example, enlarge this small box (the one with the lines bisecting each corner or what looks like an X) to fill the entire lower panel of the slide (see Figure 7.24).

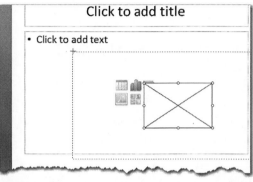

Figure 7.24

7. Right-click inside the box and select **Properties** (see Figure 7.25).

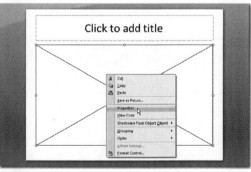

Figure 7.25

8. Under **Movie**, specify the path to your .swf file (see Figure 7.26).

9. When you are finished, close the window. Save your PowerPoint file.

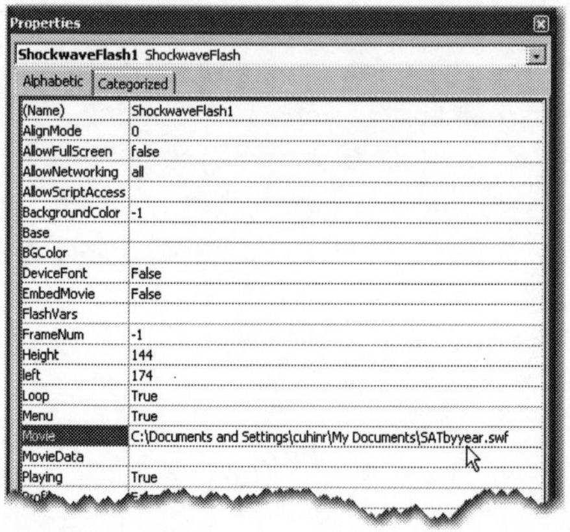

Figure 7.26

When saving this file, be sure to identify it as PowerPoint Macro-Enabled Presentation in the Save as Type drop-down menu, as shown (see Figure 7.27).

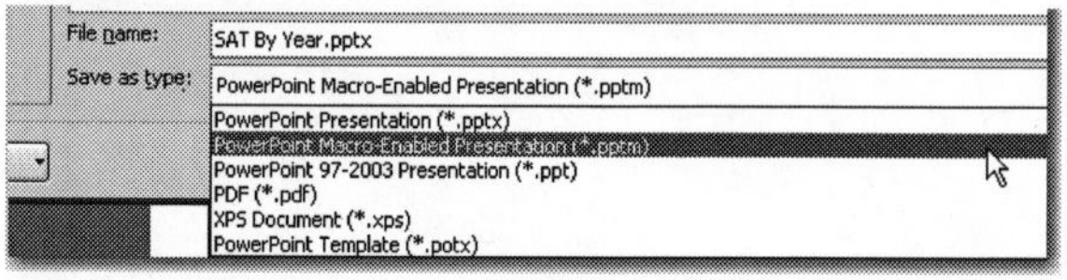

Figure 7.27

10. Select the Presentation icon (see Figure 7.28) and view the slide.

Figure 7.28   **Figure 7.29**

11. Click **Play** to see the animated graph in action (see Figure 7.29).

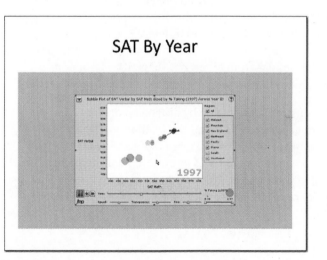

7.6 Using the Layout Option to Share Results

If you'd like to edit, annotate, or consolidate a sequence of JMP graphs or results before printing, sharing, or moving them into another application, use the Layout tool. Layout let's you create a collection of graphics or output that you can print or paste elsewhere. The usual interactive JMP graphs are fixed after being placed into Layout.

Layout allows you to ungroup the elements of graphs and results and piece together only those components you need for your presentation. For example, if you'd like to change the title, remove some pieces of an analysis report, combine multiple graphs from different JMP platforms, or add annotations to elements of multiple graphs, Layout can help.

Let's illustrate this feature by picking up on our SAT example introduced earlier. We used the Distribution platform to create histograms of ACT Score (1997) and Student/Faculty Ratio and noted that it was interesting that there might be a positive relationship between these columns (as the Student/Faculty ratio increases, so do the ACT scores) (see Figure 7.30).

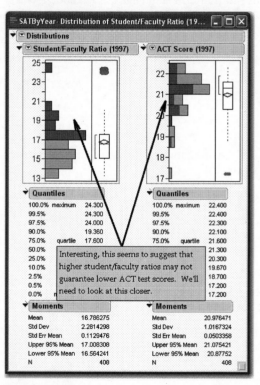

Figure 7.30

So let's answer this question by adding the appropriate additional analyses to the mix. Our first step after annotating the distribution screens is to move these graphs into the Layout platform:

1. Choose **Edit > Layout** (see Figure 7.31). A new window is created with the header Layout: Untitled. Think of this new window as a poster board that you can layout and annotate any way you like. Now you can add other graphs to this window with ease.

2. Return to the data file and select **Analyze > Fit Y by X.** Then select **Student/Faculty Ratio (1997)** for Y and **ACT Score (1997)** for X. Click **OK**. From the red triangle, select **Fit Line** and observe the graph depicted by the relationship. Does it appear from the graph that as one column increases, another also increases or decreases? We now want to combine this scatter plot with the histograms noted earlier.

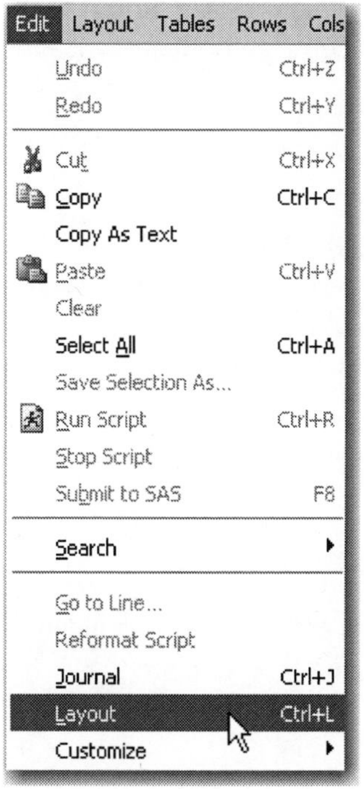

Edit	Layout	Tables	Rows	Cols

Undo	Ctrl+Z
Redo	Ctrl+Y
Cut	Ctrl+X
Copy	Ctrl+C
Copy As Text	
Paste	Ctrl+V
Clear	
Select All	Ctrl+A
Save Selection As...	
Run Script	Ctrl+R
Stop Script	
Submit to SAS	F8
Search	▶
Go to Line...	
Reformat Script	
Journal	Ctrl+J
Layout	Ctrl+L
Customize	▶

Figure 7.31

With the Selection tool, highlight the graphs and output you want to combine with the distribution graphs. Right-click and select **Copy**.

3. Go to the Layout: Untitled window, right-click, and select **Paste**.

4. With the Arrow tool, choose **Edit > Select All**. An entirely new drop-down menu, **Layout**, appears. Choosing **Layout > Ungroup** (see Figure 7.32) allows you to keep, delete, or move the elements of these two separate analyses as you see fit.

   a. Use the Selection tool to change or delete individual elements.

   b. Use the Arrow tool to move things around.

   c. Use Lines, Annotate, and other tools as you would in other platforms (see Figure 7.33).

*Note: With the **Layout** menu, you may also **Align** the contents into an efficient layout or **Group** (or regroup) separate elements together.*

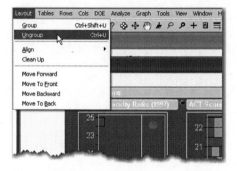

Figure 7.32

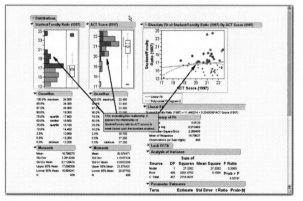

Figure 7.33

7.7 Using Scripts to Save or Share Work

Sometimes you'll find yourself exploring your data with a sequence of graphs and results and you will want to capture all of them without having to paste them individually into layout or another document. Scripts are the answer to that need. One of the things that even experienced JMP users might not realize is that JMP has a built-in scripting language that is always running behind the scene and can capture the work you've done so that you can easily reproduce it. It is always there and available by simply clicking on the menu.

From the red triangle of any platform or graph is a **Script** option. When selecting the **Script** option, you see a submenu.

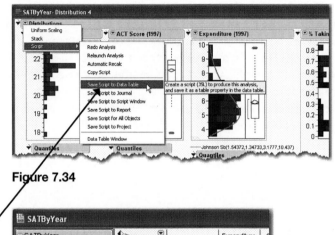

Figure 7.34

1. **Save Script to Data Table** translates this work into a script and places it in the Table panel of the Data Table (see Figure 7.34).

2. Go to the Data Table. You now see an item with the title of the platform you were using in the Table panel with a red triangle (Distribution in this case). Select the red triangle next to Distribution and **Run Script** (see Figure 7.35). You should now see exactly what you developed before. You can save and share these scripts (within the Data Table, in this case) or rename them to something more descriptive.

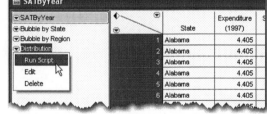

Figure 7.35

7.8 Summary

In this chapter, we learned a little bit about communicating with graphs and about sharing JMP graphs and files and moving them into other applications. We only scratched the surface in this chapter, but we tried to address the most common and basic functionality. As mentioned, there are excellent reference materials available, both within JMP and from books dedicated to these topics. For more information see the Bibliography section in this book.

Chapter 8 • Getting Help

We designed this book to be a quick overview of the most commonly used features of JMP. As you begin to use JMP more frequently or dig deeper into its analytics, you inevitably will have questions or find places where this book is insufficient. This chapter is designed to help you navigate the extensive resources available to help you answer those questions. JMP contains a wide range of built-in resources, and we begin with these.

Within JMP lives a wide variety of documentation, tutorials, and other support materials. It is highly recommended that you learn about these because they were designed to assist you while you are using the software. These tools offer the most immediate answers.

Should you find these resources insufficient, there are other resources, including outstanding technical support, webinars, user groups, blogs, and even social networking initiatives. JMP puts on dozens of events each year and also hosts its annual Discovery (user) Conference. JMP training courses are also available through SAS Education, and many books, ranging from elementary to advanced, are referenced in the bibliography at the end of this book.

8.1 The Help Tool

When you have any question about what you're seeing in a JMP window, what a graphic or result is, or how it was generated, the Help tool (also known as the question mark tool or the **?** icon) in the **Tools** menu is a great place to start. The Help tool provides a direct link to the documentation concerning the item in question.

1. Select the question mark from the toolbar and move the transformed **?** cursor to the item in question (see Figure 8.1).

2. Left-click on the item and you see that it has taken you to the appropriate place in the documentation that addresses the concepts or output at hand (see Figure 8.2).

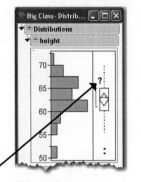

Figure 8.1

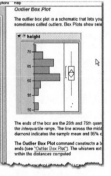

Figure 8.2

8.2 The Help Menu

From the **Help** menu (see Figure 8.3), you can access specific topics along with all of JMP's documentation, tutorials, sample data, and shortcuts to every statistical routine in JMP. Let's define each of these items in the menu.

Contents, Search, Index

Each of these first three items is designed for navigating the user documentation:

- **Contents** provides a Table of Contents view.

- **Search** provides a hypertext search that allows you to search for all the contents related to a topic or phrase.

- **Index** lists the key terms and allows you to find a topic in the user documentation.

*Note: If you are an existing user, there is a handy summary of new features to the version you are using. Select **Help > Contents > New Features in JMP 8** (see Figure 8.4).*

Figure 8.3

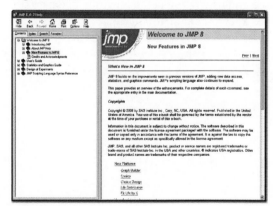

Figure 8.4

Help

Tip of the Day

JMP has summarized the most common and important FAQs into a collection of 36 *Tip of the Day* boxes (see Figure 8.5). These are common features users ask for and just haven't realized JMP already offers. Often these are somewhat hidden features but worthy of special mention in this form.

After JMP is installed, Tip of the Day boxes appear each time you launch JMP (see Figure 8.6). We encourage you to review these at least once because many of them will save you valuable time and frustration in the long run. If you do not want these to appear at launch, you can remove this default by un-checking the box at the lower left corner of the window or by selecting **File > Preferences > General** and unchecking the **Show the Tip of the Day at Startup**.

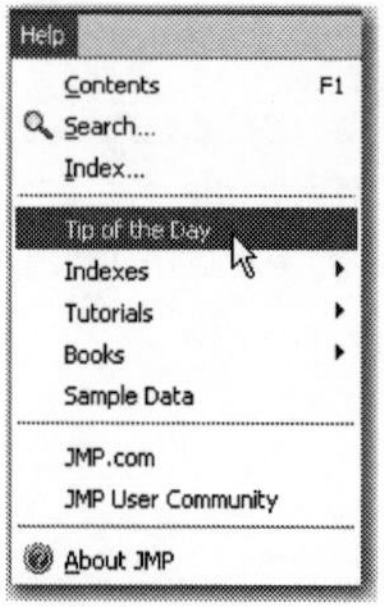

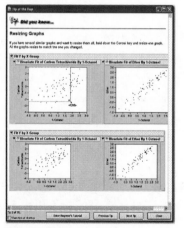

Figure 8.5

Figure 8.6

Indexes

Indexes is the next item on the Help menu. Select **Indexes > Statistics** when you know which statistical method you need but not where to find it in JMP (see Figure 8.7). **Indexes > Statistics** provides the following information:

a. A comprehensive and alphabetical index of statistical tools included in JMP.

b. Brief definitions of statistical terms with links to more comprehensive information through the **Topic Help** button. In some cases, an **Example** button is available to illustrate the technique.

c. A direct shortcut to the JMP window or platform which generates a desired statistic.* Select **Help > Indexes > Statistic > *choose statistic* > Launch** (see Figure 8.8).

Note: When using the Indexes as a shortcut, you need to have your data open first or follow the prompts to open it.

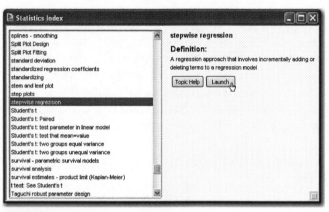

Figure 8.7

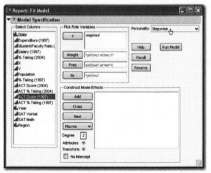

Figure 8.8

Help

Tutorials

The tutorials cover the more commonly used statistical tools in JMP (see Figure 8.9). Each of these can be quickly completed to help you not only master the JMP steps but also review the concepts involved. Each tutorial walks through the concept at hand, illustrating how to perform the analysis with JMP while explaining what each step does.

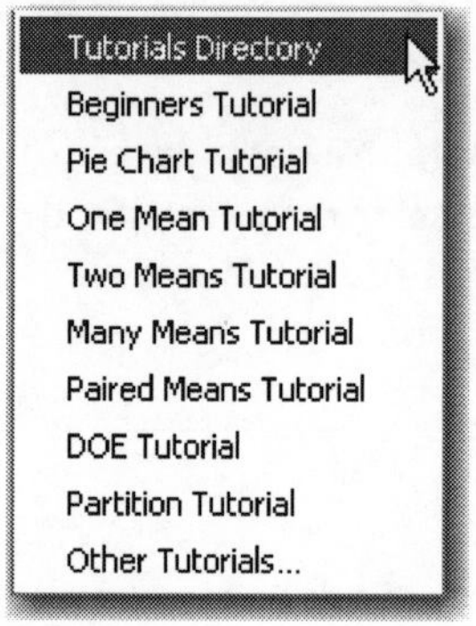

Books

This menu links to PDF versions of each of the user manuals developed for JMP (see Figure 8.10). More than 2,700 pages of JMP reference material are right at your fingertips. You can search and even print any of these manuals. Handy quick reference cards that you can print are also included. The *JMP Introductory Guide* (renamed *Discovering JMP* in JMP 9) is a logical next step for the reader of this book because it covers the basics in slightly greater depth.

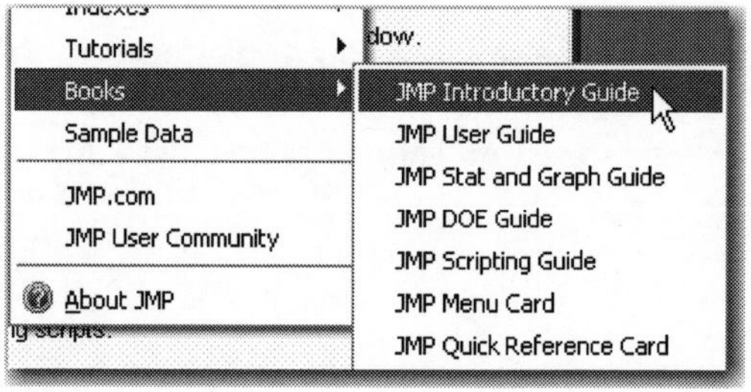

Figure 8.9

Figure 8.10

Help

Sample Data Index

The Sample Data Index contains all the data used in this book and JMP documentation (see Figure 8.11). It provides a convenient place to access data sets that you might want to employ to see how JMP performs a specific method or type of problem or to use for teaching purposes. It also contains other examples that might be of interest and provides a guide for how your data should be structured for a particular analysis.

The sample data contains over 300 data sets and includes all of those used in the books described in the previous section. Most data tables contain scripts that will generate the analysis as expressed in the JMP documentation. To generate these, simply locate the scripts in the Data Table's top left panel.

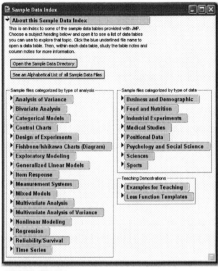

Figure 8.11

- Click on the hot spot and select **Run Scrip**t (see Figure 8.12).

*Note: The **Open the Sample Data Directory** and **See an Alphabetical List of all Sample Data Files** buttons in this window allow you to quickly locate and access a data table, including those used in this book.*

Figure 8.12

Help

8.3 The JMP Starter

The JMP Starter window provides an alternate means to navigate JMP along with a useful way to learn about the software. As discussed in Chapter 1, this book focuses on the native menus, which appear at the top of the JMP window. For some users, however, the JMP Starter is a helpful and natural interface to JMP. For these users, we cover the JMP Starter essentials in this section. The JMP Starter consists of 12 categories with corresponding topics within each category.

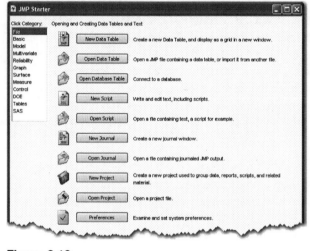

- Select **View > JMP Starter** (see Figure 8.13). Click through the items in the **Click Category** box on the left and see the corresponding topics within each category to the right. Notice that it organizes topics slightly differently and provides a description of the graphs and techniques within each option.

- Click on one of the topic buttons and a corresponding window launches to generate that output (similar to Indexes in the Help menu).

There is extensive overlap with the native menus we've been discussing to this point, but the JMP Starter provides more details. It is merely a matter of preference. If you've used similar products or you prefer this organization, we recommend you use it.

Figure 8.13

*Note: Like the Tip of the Day, the JMP Starter launches at start up as a default. If you want to change this, select **File > Preferences > General**, uncheck the **Initial JMP Starter Window**, and click **Apply** and **OK**.*

Help

8.4 External Resources

When you can't find answers within JMP's built-in resources, there are many external resources available. JMP has a very useful Web site, provides exceptional technical support, and offers many events, including the annual Discovery Conference, regional user groups, and free webinars. Comprehensive JMP training is available through SAS Education at SAS Regional Training Centers, at company sites, and through the Web. In this section, we introduce you to some of these resources.

The JMP Web Site

Think of the JMP Web site as the source for the latest information and a portal to the resources we discuss in this section. The Web site is your gateway to the **JMP User Community**, white papers, product information, and to downloads, such as maintenance upgrades. You can sign up for a webinar, learn about the latest JMP events, or read JMP Newsletters. The URL is

www.jmp.com

The JMP User Community contains

- User-developed scripts, helpful tools, and custom menu items.
- Tools for teaching with JMP.
- A JMP blog and discussion forum.
- Webcasts.

JMP Webcasts

JMP Webcasts are offered each week for new users of the software at no cost. JMP also provides less frequent webinars on more advanced features of JMP. These are live events hosted by JMP systems engineers and developers. Recorded or "On-demand" versions are also available for your convenience. Go to **http://www.jmp.com/about/events/webcasts/** for more information and to register.

JMP Technical Support

JMP Technical Support provides registered users with support from your phone and e-mail. JMP Technical Support is staffed by statisticians who not only know the software but who are often well-versed in the methodology itself.

Tech Support by e-mail: **support@jmp.com**

Tech Support by phone: 1-919-677-8008

JMP Japan Tel: +81 3 3533 3887

JMP China Tel: +86 21 61633088

JMP Germany Tel: +49 6221 / 415 - 200

You can also find many Frequently Asked Questions (FAQs) and known issues at the Knowledgebase: **http://www.jmp.com/forms/knowledge_base.shtml**.

SAS Training for JMP

SAS Education offers several courses on JMP ranging from basic to advanced methodologies. These courses are offered at SAS Training Facilities throughout the world, at customer sites, and through the Web. The courses offered at the time of publication include the following:

- Data Exploration
- ANOVA and Regression
- Introduction to Categorical Data Analysis
- Introduction to JMP Scripting Language
- Measurement System Analysis
- Process Control Design using SPC
- Stability Analysis
- Analysis of Dose Response Curves
- Classic Design of Experiments
- Custom Design of Experiments
- Mixture Design of Experiments
- Statistical Quality Control

For more information about these courses or to register, go to **http://support.sas.com/training/**.

User Groups

There are several regional user groups that have regular meetings. User groups are a convenient way to connect with other JMP users and to learn about some of the latest developments with the software. For more information, go to **http://www.jmp.com/about/events/usergroups/**.

Help

Appendix A • Integrating with SAS

A.1 Working with SAS Data
A.2 Working with SAS Programs

When you use JMP, you are using one in a family of SAS products. JMP is designed as a nimble and fast desktop data discovery tool for use by a broad audience. At the beginning, SAS software was created to provide analytics on mainframe computers when few other options were available. Over the years, as problem sizes grew, so did SAS software. Today, 98% of Fortune 500 companies use SAS software on every size and type of problem.

For several years, JMP has been an integrated client to SAS, not just a stand-alone product as described in this book. This means that as your problem size grows or extends beyond that which is suitable for JMP alone, you can migrate smoothly from desktop JMP to other tools from SAS. When analyzing SAS data with JMP you need to first import that data into JMP, which we will describe in the next section.

A.1 Working with SAS Data

SAS data files can be stored either on your local machine or on a network drive. You open them in the usual way using the **File** menu.

Select **File > Open > Files of Type** and you see that JMP can read a variety of SAS data and file types (see Figure A.1). Locate the SAS file and select **Open**.

Many organizations connect different data sources using SAS to automate and streamline data integration. If you have an individual copy of SAS or if your organization has a SAS Metadata Server, then JMP can connect to data in these SAS frameworks.

If you have a metadata sever or you have SAS installed locally, you can select **File > SAS > Browse Data** (see Figure A.2).

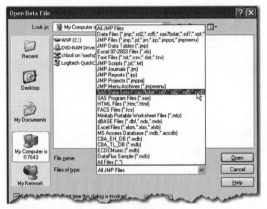

Figure A.1

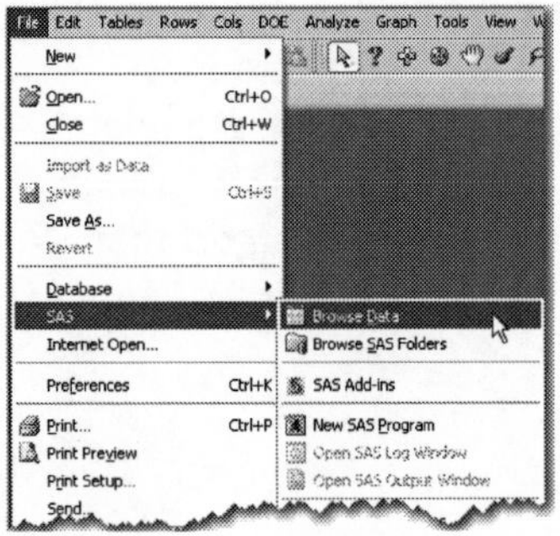

Figure A.2

You are prompted to log into SAS (see Figure A.3). This is a powerful feature of SAS that provides a permission and security framework to keep data safe.

Type your user name and password and click **OK**.

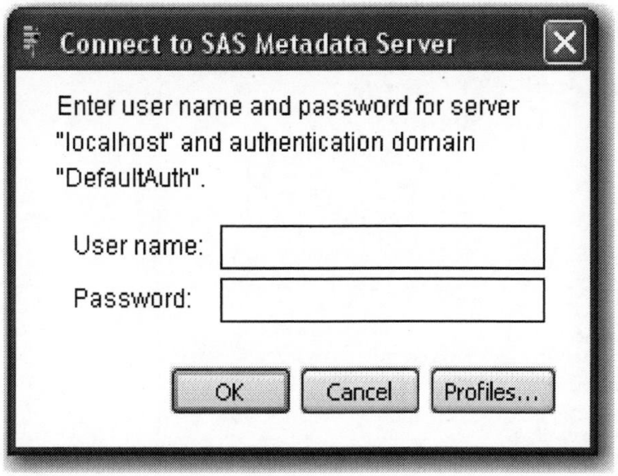

Figure A.3

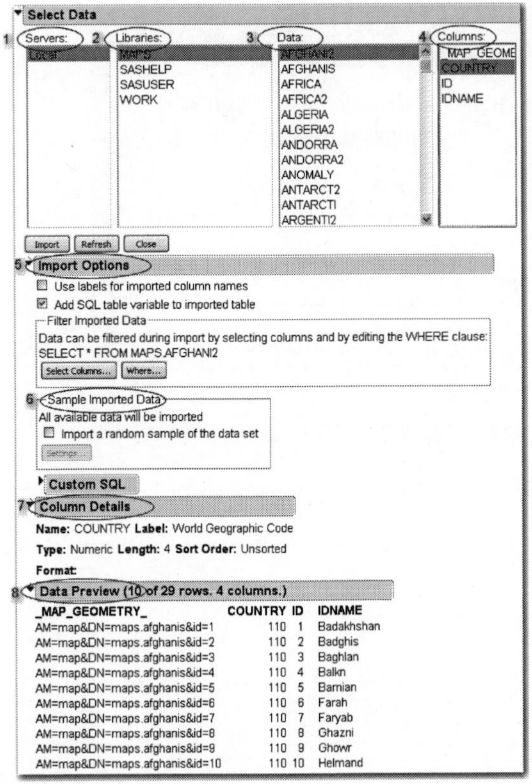

Figure A.4

There is a convenient data import window that has many useful features (see Figure A.4).

1. Local and network SAS **Servers** are shown in the Server column.

2. SAS **Libraries** that you have permission to access are shown.

3. **Data** Tables within those libraries that you have permission to access are shown.

4. **Columns** within those data tables that you have permission to access are shown.

5. An interactive **Import Options** panel allows writing of SQL queries that can be run on the data. This is supported under the **Further Import Options** report tab.

6. A **Sample Imported Data** panel allows random sampling against the data. This is a critical feature when you are importing and managing large data sizes in JMP.

7. The **Column Details** panel shows details about the column(s).

8. A **Data Preview** panel details the file size in rows and columns before it is imported to JMP. This is also valuable when accessing especially large data sources that might be too large to fit on your desktop computer.

A SAS Metadata Server also provides tools to prepare and report data using server resources that JMP can leverage. We describe a few highlights in the next section.

A.2 Working with SAS Programs

SAS supports its own programming language that scales to large and complex problems and bigger computer hardware and provides a rich set of tools.

When JMP has access to SAS, JMP can leverage these tools to extend its capabilities in several ways.

SAS Add-ins

SAS Add-ins are SAS Programs that are packaged in an easy-to-use JMP dialog window. SAS Add-ins allow you to take advantage of SAS' advanced analytics or custom applications in a convenient and accessible manner without needing to program in SAS. In this way, JMP can work as an easy-to-use client to SAS, assuming SAS is installed and connected.

From JMP, select **File > SAS > SAS Add-ins** (see Figure A.5).

Figure A.5

Appendix A

A small window of choices appears (see Figure A.6). The SAS Add-ins shown here are those that are included with JMP. Additional add-ins may be created* to generate specific SAS features and functionality.

As an example, when **Loess Regression** is selected, you are prompted to connect to SAS. After connecting, a window appears that prompts you for variable selections and parameters for the analysis (see Figure A.7).

Figure A.6

*Note: We recommend that you consult with a SAS expert if you want to create your own custom add-ins.

Figure A.7

Roles, options, and plots can be specified from tabs in the window. When you click **Run**, JMP writes and submits a SAS program and returns the output in JMP (see Figure A.8).

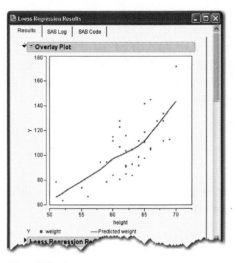

Figure A.8

Appendix A

Stored Processes

Stored processes are simply programs that run in SAS that do useful things like manage large data transformations and run customized or scheduled reports. JMP can run stored processes when you have a SAS Metadata Server. Here's how:

Select **File > SAS > Browse SAS Folders** (see Figure A.9).

You are prompted to log into SAS if you have not already done so.

Depending on your security permissions, you can view stored processes in a folder structure that might include access to locally stored processes or those supported on your SAS Metadata Server (see Figure A.10).

Stored processes can prompt you for parameters like date ranges and then run complex processes on the server that return results to JMP seamlessly. Stored processes can also be embedded in JMP menu items* (see Figure A.11). In such a menu, a stored process could access data from any data source with pre-populated queries. The menu can be customized to show anything to prompt the user; in this case, it shows a pre-programmed query to Teradata. The menu choice then returns the result to JMP on the desktop for exploration.

*See **Help > Books > JMP User Guide** for details on customizing a menu.

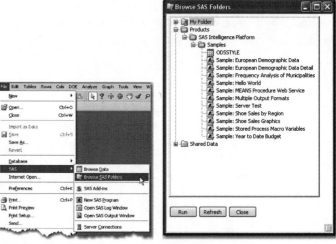

Figure A.9   **Figure A.10**

Figure A.11

Writing SAS Code in JMP and Submitting It to SAS

SAS programs can be written in JMP and submitted to SAS. In JMP, select **File > SAS > New SAS Program** (see Figure A.12).

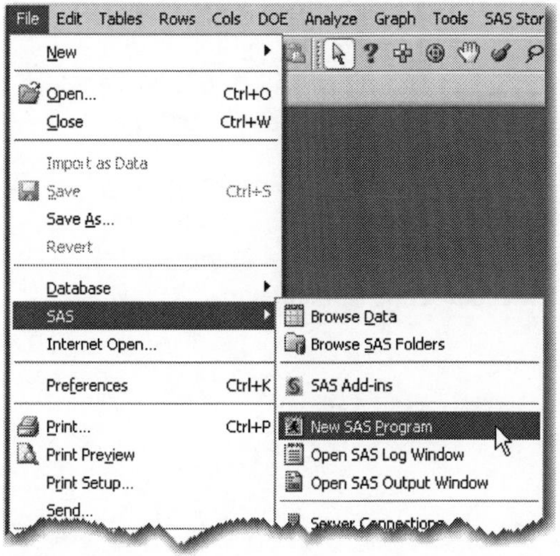

Figure A.12

A blank coding window appears. Type your program in the window. To submit the program to SAS, right-click in the coding window and select **Submit to SAS**. See Figure A.13, which contains a sample SAS program for illustration.

If you want to begin learning about SAS programming, we recommend *The Little SAS Book: A Primer* by Lora Delwiche and Susan Slaughter or *SAS for Dummies* by Stephen McDaniel and Chris Hemedinger.

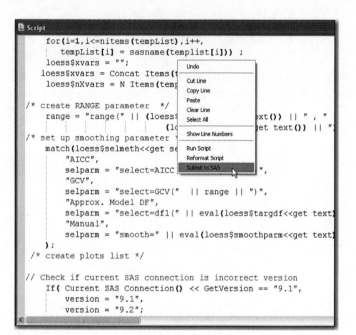

Figure A.13

Automatically Generating a SAS Program

Some limited support for automatically generating a SAS program within JMP is available. Within the Fit Model platform (and ARIMA in Time series) in JMP, SAS program code can be automatically generated and then submitted by JMP. Let's try it out:

Select **Help > Sample Data** and then select **Examples for Teaching > Big Class** (see Figure A.14).

Within JMP, select **Analyze > Fit Model**. Select **weight** for the Y role and select **height** and click **Add**. Then select the red triangle/hot spot in the upper left of the window and select **Create SAS Job** (see Figure A.15).

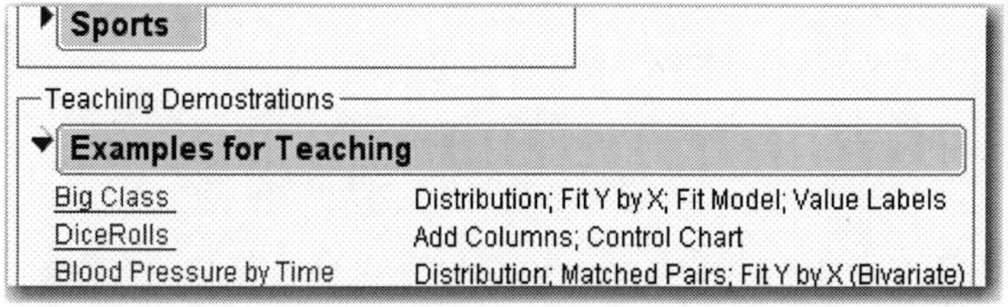

Figure A.14

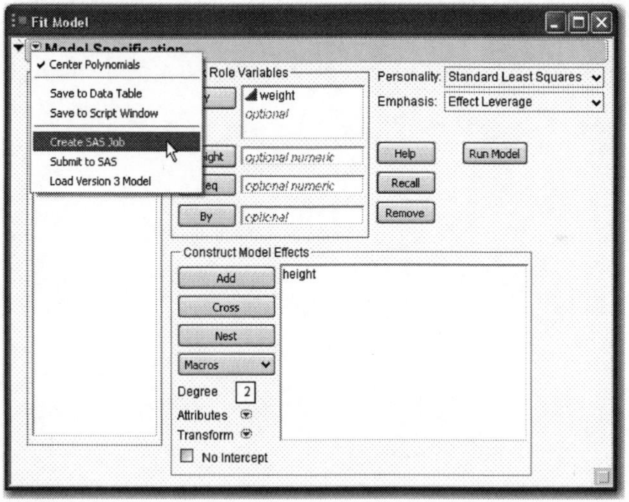

Figure A.15

This creates the SAS program in a separate window for the model you specified (see Figure A.16).

This SAS program can now be submitted to SAS by right-clicking in the window and selecting **Submit to SAS**.

The program runs in SAS and returns the results to JMP (see Figure A.17).

Of course, in this simple model JMP could easily generate the results, but when you want to run a very large or complex model, SAS can provide more options and will provide more detailed results.

Figure A.16

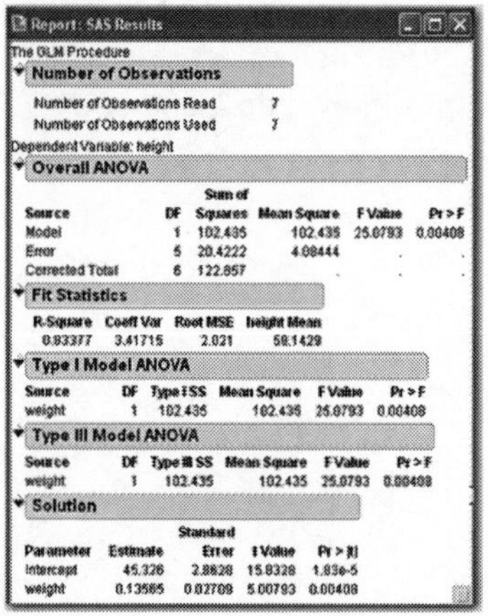

Figure A.17

Why SAS? Why JMP?

SAS is a highly scalable analytics and data management environment that can handle very small problems to massive ones. JMP is designed to be easy to use and fast, but your data tables need to fit within the memory (or RAM) available on the desktop when you are using it. JMP 8 can handle data table sizes in the 1-gigabyte range in its 32-bit version and up to the 100-gigabyte range in its 64-bit version. When problem sizes exceed these limits, SAS provides a convenient alternative. SAS uses available disk space to process data and, therefore, is not limited to the memory available on a desktop. SAS can process any data size into the terabytes and beyond.

SAS is also a very rich and robust environment for analytics and data management. When larger-sized problems appear or if you are thinking about production-type data management, security, and reporting, SAS provides the tools and means to handle these needs seamlessly.

Appendix B • Understanding Results

In the process of analyzing data, you will encounter terms in the results that reference vital information. Throughout Chapters 5 and 6, we discussed many of the relevant terms and results in the context of the application and analysis we were performing. This appendix is designed to provide a basic description of the common terms used in each of those platforms to help you understand your own results. We also point out where you can find complete descriptions in JMP's documentation. In most cases, you can simply use the "?" help tool and select the term in question to obtain more information. In other cases, by holding your mouse cursor over statistics, you can see Hover Help, which provides context-specific assistance in interpreting a statistic.

While a graph can be worth a thousand words and is more easily interpreted, the statistical results generated from the Analyze menu contain numerical results and terms that might be less transparent. It is important that you understand the basic ideas behind the results. This section offers a basic and general reference rather than a comprehensive one. To increase your confidence in interpreting statistical results, we encourage you to seek advice from an experienced data analyst or reference the books that ship with JMP (see the Help menu).

Recall that in Chapters 5 and 6, we generated results to answer these questions and these results contained some special terms and concepts. This appendix covers these key terms and related concepts, in alphabetical order. For the purpose of this appendix, the terms *column* and *variable* are interchangeable in these definitions.

Bivariate Plot: Also known as a *scatter diagram* or *scatterplot* where each point in the plot expresses both an *X* and *Y* value. These two-dimensional plots are used when comparing one continuous column with another continuous column. See **Help > Books > JMP Stat and Graph Guide > Bivariate Scatterplot and Fitting**.

Box and Whisker Plot: Also called a *box plot* or an *outlier box plot*. A graphical presentation of the important characteristics of a continuous variable. Box plots display the interquartile range of the data (the "box"), the spread (the whiskers), and potential outliers (disconnected points). Box plots are useful for describing variables that have a skewed distribution and for comparing two or more distributions.

Confidence Interval: An interval within which we expect a value to fall with a given certainty. For a 95% confidence interval, we are 95% certain, or confident, that a new value will fall within this interval.

Contingency Table: A table showing the observed frequencies of two nominal variables, with the rows indicating one variable and the columns indicating another.

The table reports a chi-square statistic. See **Help > Books > JMP Stat and Graph Guide > Contingency Table Analysis**.

Count: The total number of members in a group.

Correlation: A measure of relationship between two or more variables. It is a relationship where changes in the value of one variable are accompanied by changes in another variable(s). For example, a correlation of 1 indicates that as one variable increases, the other variable increases the same amount. A correlation of -1 would indicate that as the value of one variable increases, the other variable will decrease by the same amount. See **Help > Books > Stat and Graph Guide > Bivariate Scatterplot and Fitting**.

Degrees of Freedom: Also abbreviated DF, degrees of freedom are associated with many statistical estimates. Intuitively, degrees of freedom are the number of freely varying observations in a data-based calculation. The larger the sample size, the larger the degrees of freedom and the stronger the inferences we can draw about population parameters. See **Help > Books > JMP Stat and Graph Guide > Bivariate Scatterplot and Fitting**.

Distribution: The values of a single column or variable in terms of frequency of occurrence, spread, and shape. A distribution can be observed (based on data) or theoretical. Some examples of theoretical distributions are normal (bell-shaped), binomial, and Poisson. Distribution is also JMP's univariate platform.

F Ratio: In an ANOVA, the ratio of between-group variability to the within-group variability. The *F* ratio is used, in conjunction with Prob > *F*, to test the null hypothesis that the group means are equal (that there is no real difference between them). In general, larger *F* ratios indicate significant differences between at least two means. See **Help > Books > JMP Stat and Graph Guide > Bivariate Scatterplot and Fitting** and **Help > Books > JMP Stat and Graph Guide > Introduction to Model Fitting**.

Frequency: The number of times a categorical value is observed, a type of event occurs, or the number of elements of a sample that belong to a specified group. It is also called *count*.

Interquartile Range: A measure of variability or dispersion for a continuous column calculated as the difference or distance between the 25th and 75th percentiles (the first and third quartiles, respectively). Fifty percent of values fall within the interquartile range and is expressed by the box in a box and whisker plot.

Logistic Regression: A type of regression technique where the *Y*, dependent variable is nominal or ordinal and at least one *X*, independent variable is continuous. Logistic models (sometimes called *logit models*) are used to predict the probability of occurrence of an event based on the values of one or more continuous variables. See **Help > Books > JMP Stat and Graph Guide > Simple Logistic Regression**.

Maximum: The largest value observed in the sample.

Mean: A measure of location or central tendency of a column of continuous data. It is the arithmetic average computed by summing all the values in a column and dividing by the number of non-missing rows.

Median: A measure of location or central tendency of a continuous column of data. It is the middle value in an ordered column, which divides a distribution exactly in half. Fifty percent of the values are higher than the median and 50% are lower.

Minimum: The smallest value observed in the sample.

Multiple Regression: An analysis involving two or more independent *X* variables as predictors to estimate the value of a single dependent variable. The dependent *Y* variable is usually continuous, but the independent *X* variables can be continuous or categorical (provided that at least one of them is continuous). The model is usually estimated by the method of standard least squares. See **Help > Books > JMP Stat and Graph Guide > Introduction to Model Fitting**.

One-Way Analysis of Variance (or One-Way ANOVA): A procedure involving a categorical *X*, independent variable and a continuous *Y*, dependent variable. One-way ANOVA is used to test for differences in means among two or more independent groups (though it is typically used to test for differences among at least three groups because the two-group case can be analyzed with a *t*-test). When there are only two means to compare, the *t*-test and the *F*-test are equivalent. See **Help > Books > JMP Stat and Graph Guide > One-Way ANOVA**.

Outlier: An observation that is so extreme that it stands apart from the rest of the observations; that is, it differs so greatly from the rest of the data that it gives rise to the question of whether it is from the same population or involves measurement error. One statistical definition for an outlier is any value that is 1.5 times the interquartile range if the distribution is approximately normally distributed.

Partition: The Partition platform iteratively separates data according to a predictive relationship between a *Y* and multiple *X* values, from strongest to weakest, forming a tree structure. Partition searches through the data table to find values within *X* columns that best predict the outcome of *Y*, your column of interest. Partition is a data mining or predictive modeling technique. See **Help > Books > JMP Stat and Graph Guide > Recursive Partitioning**.

Prob > F: In ANOVA, the probability of obtaining (by chance alone) an *F*-value greater than the one calculated if, in reality, there is no difference in the population group means. Prob (or "p" values) of 0.05 or less are often considered evidence that a model fits the data. See **Help > Books > JMP Stat and Graph Guide > Bivariate Scatterplot and Fitting**.

Prob > t: A *p*-value or measure of significance for a *t*-test. For a one-sample test or for a test of differences between two means, it is the probability of obtaining a value more extreme than the hypothesized value, if the null hypothesis were true. Prob > *t* values of 0.05 or less are usually considered significant. See **Help > Books > JMP Stat and Graph Guide > Introduction to Model Fitting**.

Quantiles: Values that divide an ordered set of continuous data (from smallest to largest) into equal proportions. Related terms are *deciles* (dividing data into 10 parts) and *quartiles* (dividing data into four parts, or quarters). Values in the 97th percentile, or quantile, are equal to or larger than 97% of all values in the distribution.

Quartiles: Values in a continuous column of data that are first ordered (from smallest to largest) and then divided into four quarters, each of which contains 25% of the observed values. The 25th, 50th, and 75th percentiles are the same as the first, second, and third quartiles, respectively. See also **Quantiles**.

Regression: A statistical procedure that shows how two or more variables are related, which is represented in the simple case by a fitted line in a bivariate scatterplot. The fitted line, along with its regression equation, allows one to predict values of *Y* based on observations of *X*. The simple form of this equation is expressed as $y=mx+b$, which determines the extent to which one variable changes with another.

RSquare: A measure of the adequacy of a model defined as a proportion of variability that is accounted for by the statistical model. RSquare provides a measure of how well future outcomes are likely to be predicted by the model. See **Help > Books > JMP Stat and Graph Guide > Bivariate Scatterplot and Fitting**.

Standard Deviation: A measure of variability or dispersion of a data set calculated by taking the positive square root of the variance. It can be interpreted as the average distance of the individual observations from the mean. The standard deviation is expressed in the same units as the measurement in question. It is usually employed in conjunction with the mean to summarize a continuous column.

Standard Least Squares: A method of fitting a line to data in a bivariate plot or multiple regression model. Least squares is used where there is one continuous Y and at least one continuous X column in your model. Least squares fits a model that minimizes the total sum of squares in the data, hence the name "least squares." The least squares method is used extensively for prediction and for calculating the relationship between two or more variables. See **Help > Books > JMP Stat and Graph Guide > Introduction to Model Fitting**.

Sum of Squares: A measure of variation of the model to the observed data. This is calculated by squaring the vertical distance between each data point and the model fit (the error) and summing the values. The best model fit is one where the total sum of squares is minimized.

Appendix C • JMP Shortcuts

This appendix provides the essential menu steps to generate a desired result with JMP. JMP windows require you to specify the right data and modeling types in order to generate a result. Should you have questions about modeling types or the options displayed in the windows, refer to Chapter 2.

Some windows contain options (for example, **Charts**, **Fit Model**) that should be selected before you execute your results. These steps contain a comma (,) followed by a black inverted triangle (▼) to indicate a drop-down menu within the window for required settings.

A semicolon (;) indicates that there is a secondary step in the process that occurs only after you have executed some base results. The ▼ > symbols indicate a selection is required from a hot spot (or red triangle) drop-down menu that appears in the base results.

Some of the tasks included in this section are not covered in the book, but you can reference these tools in JMP's built-in documentation at **Help > Search**.

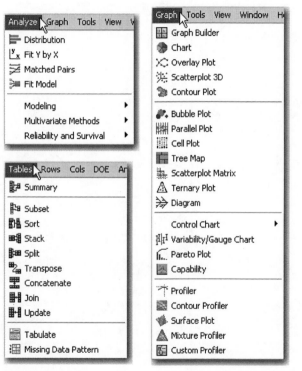

Figure C.1

Task	Menu Selection
Adding Labels	*Click on column heading;* **Cols > Label/ Unlabel**
ANOVA	
-One Way	**Analyze > Fit Y by X; ▼ > Means/ Anova**
-Two or More Factors	**Analyze > Fit Model**
Bar Chart	**Graph > Chart**
Basic Charts	**Graph > Chart**
Bivariate	**Analyze > Fit Y by X**
Box Plots	
-One Level	**Analyze > Distribution; ▼ > Outlier Box Plot**
-Two or More Levels	**Analyze > Fit Y by X; ▼ > Display Options > Box Plot**
Bubble Plot	**Graph > Bubble Plot**
C-Chart	**Graph > Control Charts > C**
Chart	**Graph > Chart, ▼** *type*
Chi-Square	**Analyze > Fit Y by X**
Color or Mark by Column	**Rows > Color or Mark by Column**

Task	Menu Selection
Column Info Dialog	*Click on column heading;* **Cols > Column Info**
Concatenate	**Tables > Concatenate**
Contingency Platform	**Analyze > Fit Y by X**
Control Charts	**Graph > Control Chart >** *Select type*
Correlation	**Analyze > Multivariate Methods > Multivariate**
Covariance	**Analyze > Multivariate Methods > Multivariate; ▼ > Covariance Matrix**
CUSUM	**Graph > Control Chart > CUSUM**
Data Filter	**Rows > Data Filter**
Data Mining	*see* Partition
Density Ellipses	
-Bivariate	**Analyze > Fit Y by X; ▼ Density Ellipses**
-Scatterplot Matrix	**Graph > Scatterplot Matrix; ▼ > Density Ellipses**
Descriptive Statistics	**Analyze > Distribution; ▼ Display Options > More Moments**
Design of Experiments	**DOE**
Distribution / Univariate	**Analyze > Distribution**

Task	Menu Selection
Distribution Fitting	**Analyze > Distribution; ▼ > Continuous Fit** (*select distribution* or "All") **or Discrete Fit** (*select distribution*)
Excel Files, to Open	**File > Open >** *Specify Excel file format*
Experimental Design	*see* Design of Experiments
Exponential Smoothing	**Analyze > Modeling > Time Series; ▼ > Smoothing Model**
Fit Line	**Analyze > Fit Y by X; ▼ > Fit Line**
Fit Polynomial	**Analyze > Fit Y by X; ▼ > Fit Polynomial** (*specify degree*)
Fit Y by X Platform	**Analyze > Fit Y by X**
Forecasting/Time Series	**Analyze > Modeling > Time Series**
Formula Editor	*Click on column heading;* **Cols > Formula** (or to create new column) **Cols > New; ▼ Column Properties > Formula**
Frequency Distribution	**Analyze > Distribution**
Full Factorial Design	**DOE > Full Factorial Design**
Gauge Chart	**Graph > Variability/Gauge Chart; ▼ > Gauge Studies > Gauge RR**

Task	Menu Selection
Goodness-of-Fit (Test for Normality)	**Analyze > Distribution; ▼ Continuous Fit > Normal; ▼** (*under* **Fitted Normal**) **> Goodness of Fit**
Graphs	
-Chart	**Graph > Chart**
-Graph Builder	**Graph > Graph Builder**
-Overlay Plot	**Graph > Overlay**
-Control Chart	**Graph > Control Chart >** *type*
-Bubble Plot	**Graph > Bubble Plot**
-Scatterplot	**Analyze > Fit Y by X**
-Scatterplot 3D	**Graph > Scatterplot 3D**
Graph Builder Platform	**Graph> Graph Builder**
Histograms	**Analyze > Distribution**
Histogram Color	*right-click on histogram > select* **"Histogram Color"**
Holt-Winters	**Analyze > Modeling > Time Series; ▼ > Smoothing Model > Winters Method**
Horizontal Bar Chart	**Graph > Chart, ▼ Horizontal**
IR Chart	**Graph > Control Chart > IR Chart**

Task	Menu Selection
Joining	**Tables > Join**
Kruskal-Wallis Test	**Analyze > Fit Y by X; ▼ > Nonparametric > Wilcoxon Test**
Least Squares Regression	*see* Regression, Simple or Multiple
Line Chart	**Graph > Chart, ▼ Line Chart**
Logistic Regression	*see* Regression, Logistic
Moments	**Analyze > Distribution**
Mosaic Plot	**Analyze > Distribution; ▼ > Mosaic Plot**
Mosaic Plot with Two Columns	**Analyze > Fit Y by X**
Moving Averages	**Analyze > Modeling > Time Series; ▼ > ARIMA**
Moving Range Chart	**Graph > Control Chart > IR,** *check* **Moving Range (Average)** *box*
Multiple Comparisons	**Analyze > Fit Y by X; ▼> Compare Means** (*specify comparison*)
Multiple Regression	*see* Regression, Multiple
Multivariate Platform	**Analyze > Multivariate Methods**

Task	Menu Selection
Oneway ANOVA	**Analyze > Fit Y by X; ▼> Means/ Anova/Pooled t**
Outlier Box Plot	*see* Box Plot
Overlay Plot	**Graph > Overlay**
p Chart	**Graph > Control Charts > P**
Parallel Plot	**Graph > Parallel Plot**
Pareto Diagram	**Graph > Pareto Plot**
Partition	**Analyze > Modeling > Partition**
Phase Chart	**Graph > Control Chart >** *select type, specify column in* **Phase** *role*
Pie Chart	**Graph > Chart, ▼ Pie Chart**
Pivot-Table	*see* Tabulate
Point Chart	**Graph > Chart, ▼ Point Chart**
Power Calculations	**DOE > Sample Size and Power**
Prediction Profiler	**Analyze > Fit Model; ▼ > Factor Profiling > Profiler**
Predictive Modeling	*see* Partition, Profiler, Fit Model
Profiler	*see* Prediction Profiler
Process Control	*see* Control Charts

Task	Menu Selection
R-Chart	**Graph > Control Chart > R**
Recursive Partitioning	see Partition
Regression, Simple -with Line Fit	**Analyze > Fit Y by X; ▼ > Fit Line**
-with Polynomial Fit	**Analyze > Fit Y by X; ▼ > Fit Polynomial**
-with Spline Fit	**Analyze > Fit Y by X; ▼ > Fit Spline**
Regression, Logistic	see Regression, Simple or Multiple
Regression, Multiple	**Analyze > Fit Model**
Regression Trees	see Partition
Residual Analysis	**Analyze > Fit Model; ▼ > Row Diagnostics > *select type***
Response Surface	**DOE > Response Surface Design**
Run Charts	**Graph > Control Chart > Run Chart**
Sample Size Calculations	**DOE > Sample Size and Power**
S-Chart	**Graphs > Control Charts > x-bar, check "S" box**
Sample Data Directory	**Help > Sample Data**

Task	Menu Selection
Scatterplot	**Analyze > Fit Y by X**
-3D Scatterplot	**Graph > Scatterplot 3D**
-Matrix	**Graph > Scatterplot Matrix**
-with Line	**Analyze > Fit Y by X; ▼ > Fit Line**
-with Polynomial Fit	**Analyze > Fit Y by X; ▼ > Fit Polynomial**
-with Spline Fit	**Analyze > Fit Y by X; ▼ > Fit Spline**
Screening Design	**DOE > Screening Design**
Simple Regression	see Regression, Simple
Sort	**Tables > Sort**
Spearman's Rho	**Analyze > Multivariate Methods > Multivariate; ▼ > Nonparametric Correlations > Spearman's *p***
Stem-and-Leaf	**Analyze > Distribution; ▼ > Stem and Leaf**
Stepwise Regression	**Analyze > Fit Model, ▼ Stepwise**
Subset	**Tables > Subset**
Summary	**Tables > Summary**

Task	Menu Selection
t- or *z*-Test	
-One Sample	**Analyze > Distribution; ▼ > Test Mean**
-Two Sample	**Analyze > Fit Y by X; ▼ Means/Anova/Pooled t**
-Paired *t*-Test	**Analyze > Matched Pairs**
Tables	
-Sort	**Tables > Sort**
-Subset	**Tables > Subset**
-Join	**Tables > Join**
-Concatenate	**Tables > Concatenate**
-Summary	**Tables > Summary**
Tabulate Platform	**Tables > Tabulate**
Test for Equal/Unequal Variances	**Analyze > Fit Y by X; ▼ > Unequal Variances**
Test for Proportions	
-One Proportion	**Analyze > Distribution; ▼ > Test Probabilities**
-Two Proportions	**Analyze > Fit Y by X**

Task	Menu Selection
Test for Normality (Goodness-of-Fit)	**Analyze > Distribution; ▼ > Continuous Fit > Normal; ▼(*under* Fitted Normal) > Goodness of Fit**
Time Series Plot	**Analyze > Modeling > Time Series**
Tree Map	**Graph > Tree Map**
Two or More Factor ANOVA	**Analyze > Fit Model**
Tukey Box Plot	*see* Box Plot
U-Chart	**Graph > Control Chart > U**
Univariate/Distribution Platform	**Analyze > Distribution**
Variability Chart	**Graph > Variability/Gauge Chart**
Wilcoxon Rank Sum Test	**Analyze > Fit Y by X; ▼ > Nonparametric > Wilcoxon Test**
Wilcoxon Signed Rank Test	**Analyze > Distribution; ▼ > Test Mean, *check* Wilcoxon Signed Rank box**
XBAR Chart	**Graph > Control Chart > XBAR**

Bibliography

Delwiche, Lora D., and Susan J. Slaughter. 2008. *The Little SAS Book: A Primer, Fourth Edition*. Cary, NC: SAS Institute, Inc.

DeVeaux, R., P. Velleman, and D. Bock. 2009. *Intro Stats 3rd Edition*. Boston, MA: Pearson Addison-Wesley.

Few, S. 2004. *Show Me the Numbers: Designing Tables and Graphs to Enlighten*. Oakland, CA: Analytics Press.

Few, S. 2009.*Now You See It: Simple Visualization Techniques for Quantitative Analysis*. Oakland, CA: Analytics Press.

Freund, R., R. Littell, and L. Creighton. 2003. *Regression Using JMP*. New York: John Wiley & Sons, Inc.

McDaniel, Stephen, and Chris Hemedinger. 2010. *SAS For Dummies, Second Edition*. Hoboken, NJ: John Wiley & Sons, Inc.

Montgomery, D. 2008. *Design and Analysis of Experiments, 7th Edition*. Hoboken, NJ: John Wiley & Sons, Inc.

Moore, D., and G. McCabe. 2005. *Introduction to the Practice of Statistics, 5th Edition*. New York: W. H. Freeman.

Peck, R., C. Olsen, and J. Devore. 2009. *Introduction to Statistics and Data Analysis, 3rd Edition*. Belmont, CA: Cengage Learning.

Reynolds, G. 2008. *Presentation Zen: Simple Ideas on Presentation Design and Delivery*. Berkeley, CA: New Riders Press.

Sahai, H., and A. Khurshid. 2002. *Pocket Dictionary of Statistics*. New York: McGraw-Hill.

Sall, J., L. Creighton, and A. Lehman. 2007. *JMP Start Statistics 4th Edition*. Cary, NC: SAS Institute Inc.

SAS Institute Inc. 2009. *JMP Scripting Guide*. Available at **Help > Books > JMP Scripting Guide**.

SAS Institute Inc. 2009. *JMP Statistics and Graphics Guide.* Available at **Help > Books > JMP Stat and Graph Guide**.

Sharpe, Norean, Richard DeVeaux, and Paul Velleman. 2010. Business Statistics. Boston, MA: Pearson Addison-Wesley.

Stine, R., and D. Foster. 2011. Statistics for Business: Decision Making and Analysis. Boston, MA: Pearson Addison-Wesley.

Tufte, E. R. 1983. *The Visual Display of Quantitative Information*. Cheshire, CT: Graphics Press.

Utts, J. 2008. *Seeing Through Statistics, 3rd Edition*. Belmont, CA: Cengage Learning.

Index

Index

Index

Index